CHAPTER 1

SCIENCE

Science,
the Universe and God

The Search for Truth

By

Keith Mayes

ISBN: 1-4140-0739-6 (e-book)
ISBN: 1-4140-0738-8 (Paperback)

This book is printed on acid free paper.

1stBooks – rev. 12/08/03

Acknowledgements

Thanks go to my older brother Malcolm for never believing a single thing I told him. You'll never know how much that encouraged me to finish this book.

Thanks also go to my younger sister Jacqueline, and of course my mother, for having the kindness to once in a while ask me something.

To my two lovely daughters, Rachel and Rebecca, I love you both, even though when I proudly showed you Halley's comet through my new telescope you both scornfully said "Is that it? That small fuzzy blob?"

Thanks to my good friend Karin McDonald for laboriously and carefully checking my manuscript for errors, and finding many.

Front cover picture; The Pleiades Open Star Cluster. By the author.

Dedication

This book is dedicated to my wife, Elizabeth, for her - very nearly - endless patience when I spent time on the computer night after night, until the wee small hours. "Yes Liz, almost finished, I'm coming to bed soon." (and the cheque's in the post!)

Thank you for suggesting I should write a book.

Thank you for all your encouragement.

Liz reckons I love our dog the most, a lovely Border Collie cross called Sox, the iMac a close second, and then her. She's wrong of course, I love the dog and the computer equally.

"As being is to becoming, so is truth to belief. If then, Socrates, amid the many opinions about the gods and the generation of the universe, we are not able to give notions which are altogether and in every respect exact and consistent with one another, do not be surprised. Enough if we deduce probabilities as likely as any others; for we must remember that I who am the speaker and you who are the judges are only mortal men."
Plato

Contents

Preface

Our universe is a truly amazing creation; beautiful, mysterious and full of endless wonders. It is so amazing that it can appear at times to be impossible to exist, yet it does; or so we believe. How and why it exists has been the driving force behind centuries of research, but the universe may ultimately prove to be impossible to understand. This does not however, prevent us from be being able to marvel at the sheer beauty of its existence.

For as long as I can remember I have wondered why things are the way they are. Why does the universe exist? How big is it? Why do we exist? Do you see the blue of the sky the same colour as I do? Does God exist? The list is endless, there are countless questions. It even puzzles me that other people are not intrigued by these things to the same extent that I am; how can they possibly not be when their own existence is one of the great marvels of the universe?

I wondered if others felt the same, and to this end published a web site titled "Theories with Problems". To my amazement the emails soon started to pour in. It transpired that there are a great many people who share my questions on 'Life, the Universe and Everything'. Many people are seeking answers to the same questions. Questions are plentiful, meaningful answers however, are a scarce commodity.

My hobby is astronomy, and has been for over 40 years. I have studied the subject since I was old enough to join the local library, and have added to it over the years with a study of cosmology and physics, covering a spectrum of subjects ranging from particle physics and quantum theory to Relativity and the Big Bang. I have gained some knowledge in my search for answers, and found many surprises. After a lifetime's conviction in the ability of science to supply answers to our questions, I have made a startling discovery. I have discovered that all we know, every fact that we hold as truth, every theory postulated, is based in its very conception, on no more than a belief. This came as a complete and totally unexpected revelation.

The purpose of this book is not to supply answers, it is to raise questions, questions that will cause the reader to wonder, and to ask "Why is that?" Asking the question is the first step on the road to understanding the answer. I do not propose any theories of my own, there are enough idiosyncratic and egotistically induced theories as it is. I do though express my opinion, and I am sure you will have yours.

My philosophy regarding knowledge is a simple one and is this: *It takes a wise man to ask a meaningful question, but any fool can give an answer.* This book is full of questions, does that make me a wise man, or just a fool with questions?

Not all the topics covered are deep, complex and meaningful, some are just odd thoughts that have popped into my head, some are nothing more than idle speculation, and others simply things that either amuse or annoy me.

My website is www.thekeyboard.org.uk "Theories with Problems". You are invited to visit my site and take the opportunity to leave a comment on a choice of message boards, participate in a number of polls, engage in a lively debate, or raise a question. Or you can just read it. Also you may contact me by email at keithmayes123@aol.com with any points in this book that you may wish to raise.

Keith Mayes

Introduction

What is the meaning of Life, the Universe and Everything?

The meaning of Life, the Universe and Everything, is perhaps the most meaningful question we can ask. The question mankind has pondered over since the dawn of time. The question that goes to the very heart of our existence. Why are we here? Why does the universe exist?
What is the meaning of it all?

When we contemplate the unimaginable vastness of the universe, the incredible diversity and complexity of life on Earth, the sheer tenacity of life to survive, the wonderful beauty of nature, we are filled with a sense of amazement. Thirteen billion years or so in the making, and here we are.
And we wonder why.

In this unbelievable universe, with its 100 billion galaxies, each containing 100's of billions of stars, we inhabit one small world circling one ordinary star. One star amongst tens of thousands of billions upon billions.
And we wonder why.

We consider the processes that followed the Big Bang. How the universe cooled as it expanded, how matter formed out of radiation, how gravity shaped star formation and galaxies, how stars created heavier elements, how those elements formed planets, and how life developed on at least one of those planets.
And we wonder why.

We study the emergence of life on Earth, how asteroid impacts, volcanoes, earthquakes, floods, fire, droughts, disease, starvation, ice ages, all threaten to extinguish that life. But it survives.

And we wonder why.

This unimaginably vast universe is but a single complex organism. Every star, galaxy, planet, atom, person, rock, black hole, asteroid, comet, are all inter-linked parts of the one same universe. We are the universe. Why are we here? What is the purpose of it all?

In order to attempt to find answers to these questions, I invite you, the reader, to join me on a journey. I have provided the questions, but you will have to arrive at your own conclusions. I hope you find the journey every bit as exciting, interesting, and amazing as I do.

01. Does infinity exist?

"To see the World in a Grain of Sand And a heaven in a Wild Flower Hold Infinity in the palm of your hand And Eternity in an hour"
William Blake *Auguries of Innocence*

The concept of infinity is a puzzling one. For example: imagine a standard pack of playing cards that consists of just one of each card but two jokers. Imagine that the packs of playing cards are infinite in number (a thought exercise only of course). We therefore have more jokers than any other card in each pack, so do we have more jokers in total? You could reply that as the packs are infinite in number they cannot be counted so it would be impossible to know. However, as the ratio of jokers to other cards in each pack is fixed, at any number of packs there will always be more jokers. This would appear to indicate, that mathematically, we can have degrees of infinity. Sounds odd doesn't it to think that we can have more infinite jokers than infinite playing cards? It is a valid mathematical argument though.

There are two fundamental problems with the concept of infinity, as I see it, and they are:
1) Is it possible for the human mind to truly grasp the concept of infinity?
2) Does infinity exist outside of theoretical mathematics?

We will first examine the problem of whether or not it is possible to grasp the concept of infinity, leaving aside for the moment the question of its actual existence.

The definition of infinity is relatively straight forward and easy to understand. We are able to accept that it describes a thing as being *'without end or limit, that which is not only without determinate bounds, but also which cannot possibly admit of bound or limit.'* We can all understand the idea of infinity as describing something that has no limit, it goes on forever. However, understanding the definition and understanding the reality are two very different things.

To try and understand the reality of infinity imagine the universe as being infinite, and imagine you are aboard a light-speed space ship and you set off on a journey towards the edge of the universe. After ten million years of travelling at the speed of light (you will not have aged at all, but in all probability would be getting a little restless by now) how far have you travelled towards the edge of the universe? The answer is of course you are not one inch closer than you were before you started your journey. The reason is because we defined the universe as being infinite. An infinite universe could have no edge as it would go on forever, therefore without having an edge it is impossible to get any nearer to it. The problem this causes with our perception of the universe is that although we may say that we understand we can get no nearer to its edge, it not having one, we simply can't help thinking that we must have made some sort of progress towards it, because we have travelled at light speed 'towards it' for ten million years. The concept that we need to get into our heads is that it is not possible to approach something that does not exist. If the universe is infinite then it has no end, no edge, so it is impossible to approach it. Are you able to imagine a journey that at the speed of light (186,000 miles per second) could last for all eternity, yet take you no closer the the edge of the universe? Are you able to imagine a universe without an edge?

Infinity is very difficult, if not impossible, for us to comprehend because everything that we regard as being real we can measure and can calculate its size. We can visualise a straight rod for example as having a length of a yard, or a mile, or a hundred miles, and so on. But how can we possibly imagine a rod that has no end or beginning? There would not be a starting place to measure it from as it stretched towards infinity; if it had a start, that you could grasp in your hand and begin to measure from, it could not be infinite, for that would be and 'end' coming from the opposite direction. Our infinite rod would have to extend infinitely in length in both directions, there can be no ends, or a middle. A circle may also be described as having no ends or a middle, but our infinite rod is very different, its ends will never meet, it is going in a straight line.

We would also need to consider the implications of having our imaginary rod of infinite length, as it would of course require an infinite universe to contain it, and that would mean that nothing else could exist other than the universe that houses our infinite rod. The universe, being infinite, could have no limit, or edge, so there could not be an 'outside' to the universe for anything else to exist in. The universe housing our infinite rod would be all there was, nothing else could exist, there being nothing else but the universe.

We can also look at the idea of having infinite regression in size, for if our universe is infinite in 'largeness' it would by necessity have to be infinite in 'smallness'. This is the same situation as we would have with our rod of infinite length that cannot have an end to it in either 'direction'. For the universe to be infinite in size it also cannot be allowed to have an end in any 'direction' which means 'smallness' as well as 'largeness'. Atoms would be divisible into particles and quarks, as indeed they are, but would then have to continue in division of size to infinity. Somehow I find this even harder to accept than infinite 'largeness'!

By using our example of an imaginary rod of infinite length we arrive at the conclusion that only one universe could exist, the one housing our imaginary rod. We would therefore be forced to discard any theories that allow for the existence of other universes. These would include theories such as our universe having been created by another universe, and that our universe may itself create a universe. Our imaginary rod of infinite length would not allow for any of this to happen. We can then go on to say with confidence that if our universe were to contain any object of infinite size, nothing else but our universe could possibly exist. That's quite a statement, so I will ask you again. Can you grasp the concept of infinity? Can you imagine something of infinite size? I cannot, I have tried but I simply cannot manage to do it. However, as I am not the sole representative of the human race (thank God!) I am unable to answer the original question 'is it possible for the human mind to grasp the concept of infinity?' No one can answer that question, you will need to decide for yourself if you can.

Having made no progress at all on the first problem - life can be so cruel - let's look at the second one, 'does infinity exist outside of theoretical mathematics?' and see if we can get anywhere this time.

The terms 'infinite' and 'infinity' appear in many theories, and form an essential part of those theories, without them the theories would collapse. The terms are very real to the theory, but can anything real, not theoretical, be infinite?

Naturally in mathematics we can have infinity, numbers go on for ever, but numbers are not real, they are abstract. We know for example that when we divide 10 by 3 the answer is 3.3333 recurring. The word 'recurring' in this instance meaning the digit 3 will repeat infinitely.

What do we have that contains infinity outside of mathematics? We have, for example, a theory for black holes that describes infinite density in the singularity at its centre, but what does that mean in the real world? Exactly what is infinite density other than an unresolvable equation that occurs in mathematics? Taking a rather simplistic view it could be argued that if one black hole has infinite density then nothing else can have ANY density. Clearly though, in this sense, we can have lots of infinite density, so the term obviously carries a meaning in mathematics that it does not have outside of it. Is the term used in the theory only because that is the way the sums work out, regardless as to its significance in the real world, or is it real?

Strictly speaking, according to Einstein's Theory of Relativity, a singularity does not contain anything that is actually infinite, only things that move (mathematically) towards infinity. This may make more sense when considering how a black hole is created. A black hole is formed when large stars collapse and their mass has been compressed down to such a small size that the powerful gravitational field so formed prevents anything, even light, from escaping from it. A black hole therefore forms a singularity at its centre from the concentrated mass of the collapsed star itself, and from the accumulated mass that is sucked into it from the surrounding area within its gravitational grasp. A singularity's mass is therefore finite, the 'infinity' can refer only to the mathematics.

Can we have an infinite universe for example? The answer is no, the universe is believed to be finite. Stephen Hawking in 'A Brief History of Time' (1989 page 44) describes the universe as being "finite but unbounded". The simplest explanation is that as the universe is known to be expanding, it cannot possibly be infinite. I need to deviate here to avoid confusion. The dictionary definition of universe is "all that is. The whole system of things." In this sense the universe is not expanding into anything other than itself, for whatever it is expanding into is part of the universe, there being nothing else but the universe. However, for the sake of simplicity, I am referring only to our Big Bang expanding universe as 'the universe'. Even if you happen to disagree with the Big Bang theory, the term 'universe' will still have the same meaning here, as it refers to 'our' universe only, and does not include whatever may or may not exist outside of it. To get back on track, I was explaining that the universe cannot be infinite because it is expanding. I realise that not everybody is in agreement with the idea of our universe being finite, many are convinced it must be infinite, so I will attempt to justify this claim.

A good place to start is to understand the very real difference between infinity and a large number.

For example, imagine a large diamond, weighing an ounce. Now imagine a super-being armed with super-tweezers, picking out atoms from this diamond one at a time, one every second, since the creation of the universe, some 13 billion years ago. How much of the diamond would by now have been removed? The answer is you could not tell without looking through an electron microscope, less than a millionth of the atoms would have been removed. Try and imagine how many atoms there are in that diamond.   Now try and imagine how many atoms there are in the entire universe.   It is a very large number, but it is finite, and is 10 followed by 80 zeros, (maybe a few more zeros, maybe a few less, it doesn't matter for this example) and can be expressed as 10 to the 80th (10^{80}).   If you want to see what it looks like:

100,000,000,000,000,000,000,000,000,000,000,000,00 0,000,000,000,000,000,000,000,000,000,000,000,000, 000.

Or written as - One hundred million, billion, billion, billion, billion, billion, billion, billion, billion

Even this very large number would count as **nothing** when compared to infinity, because infinity is not a large number, be absolutely clear on this point, **infinity is not a large number**, infinity is all there is, it is not a number. You could keep counting (or measuring) for ever, and never reach infinity, it is only a description.   Infinity describes a thing as having no end, no limit, no boundary or edge, it literally goes on forever, *ad infinitum.*

If you remember nothing else at all from this section on infinity other than infinity is not a large number, but is only a description, then you will have taken a huge stride forward in your understanding of infinity.

Because infinity is not a number, large numbers are no 'nearer' to infinity than small numbers.   Number 1 billion for example is no nearer to infinity than number 1, because the two, numbers and infinity, are in no way related.   It is then impossible to approach infinity, a thing is either infinite and immeasurable, or finite and measurable, it cannot be part way towards infinity.   Imagine running up a 'down' escalator, never moving forward.   If you run for a week you are no nearer reaching the end of the escalator than if you run for a minute, you cannot get any closer to something that has no end.

An infinite universe would exist in every direction forever, there could be nothing else, only the universe.   It is then very easy to understand now why our universe cannot be infinite, it is because it is expanding.   It cannot be both infinite and expanding.   It could be infinite OR expanding, but cannot possibly be both, that is a contradiction in terms, and we do know that it is expanding. For an explanation of the Big Bang and why we believe the universe is expanding see "Is the Big Bang Theory correct?".

I understand that many people have a problem with the idea of our universe being finite, that it has an 'end' to it, a boundary. They ask what this boundary would be physically like, as though it were some form of partition that we couldn't get through. However, there is not a particular direction that we could set off in our warp speed space craft that would lead us to a boundary, no matter how far or fast we travelled. The explanation for this seeming impossibility is that space-time is curved, thus you would be travelling in a circle that only appears to be a straight line. If, for example, it were possible to direct a laser beam from here through the centre of the universe it would not hit the other side of the universe, it would eventually hit the back of your head (metaphorically speaking). Einstein demonstrated how matter in the universe distorts the space-time continuum by accurately predicting the distortion caused by the Sun. He used a total eclipse of the Sun, being the only time that stars and the Sun can be seen at the same time in close proximity, to demonstrate that a star that was behind the Sun would in fact be visible due to the distortion of space by the Sun. As Einstein predicted, the Sun curved the light from the star around the Sun making the star visible. Strictly speaking, the Sun does not actually curve the light around itself, the entire space-time continuum is curved, the light is still travelling in a 'straight' line.

Galaxies naturally create even larger distortions, and the total mass of the universe gives a distortion that results in our 'straight' line of light curving forever through the universe and never reaching the 'end'. That's why Hawking's describes it as 'finite but unbounded'. As an aid to visualisation, but not an accurate representation, consider an ant crawling around a huge beach ball and never coming to the end, it would consider the beach ball as infinite as it has no boundaries. If you now consider the ant as only a two dimensional creature and crawling round a three dimensional beach ball, you can understand why the ant would consider the beach ball to be infinite, the three dimensional picture, that shows how restricted its movement really is, is simply not available to it. The universe, from our perspective, restricted to our view from within the universe, appears to be infinite, but this is just an illusion, we are confined to the limits of our universe and cannot escape from it. We are bounded within a finite universe.

The next question that people naturally follow up with is to ask what it is that our finite universe is expanding into. This is a good question and one that can never be answered, we will never be able to escape the confines of our universe to find out. We can only theorise about this, and there are plenty of theories to choose from. I tend to think that we are expanding into an infinite nothing, but for a fuller description see "Where did the universe come from?" But the truth will never be known, your guess is as good as mine. So what is there within our universe that we can truly apply the term infinity to? The universe itself is finite. Infinite mass, in black holes for example, would only appear to be a mathematical description. The age of the universe is finite, and even the number of particles in it is finite.

It must also be understood that in order for the universe to be infinite it must have always been infinite, right from the moment of its creation. It cannot start as a finite size and grow into an infinite size, it is simply not possible for an object to expand to infinity. How can an expanding object reach infinity? How can it stop expanding and be infinite?

Before leaving our hypothetical infinite universe, let's consider one more consequence of infinity. If the universe were infinite then EVERY possibility of everything will exist somewhere. For example, we would (not could or maybe, but definitely would) find a planet identical to ours in every way, right down to the very last detail, including such things as the pyramids in the desert and the chewing gum under the park bench. We would also find another planet just the same except perhaps for the chewing gum being an inch to the left, or one of the pyramids an inch taller. Another identical planet, but this time the world's population is exactly the same as here, except that on this planet you have a different colour car. We can then start looking for the other planets where we will find you owning the same car but of every possible colour, all other details of the planets being identical to this one. After that we can start looking for identical planets but with you owning every make of car... I think you have got the picture. In an infinite universe every conceivable possibility will exist. Ridiculous? Yes, I think so to, but that is the implication of an infinite universe. Want to find a planet identical to ours but with Great Britain nearer to the equator so it's warmer? No problem! Just keep looking, its' got to be out there somewhere, or would be in an infinite universe.

I tend to think infinity exists only as a means of description, such as found in mathematics for example, or any other thing that exists only in the abstract. I do not believe that it has any real existence in the universe such as infinite mass or infinite size. The word 'infinity' is a descriptive term and not a measure of size, and I therefore do not see how it can be applied to anything 'real', as all real things can be measured.

I have come across theories, and maybe you have also, that claim that atoms can be subdivided down into infinity, and even that they contain tiny universes within them, and no doubt tiny people as well. Although science has not yet been able to prove we have reached the ultimate elementary particle from which all complexity is built, there is very strong theoretical and experimental evidence to show that quarks could be it. Smaller than quarks enters the realm of energy, not particles, as in string theory. As matter has been subdivided down from complex objects, to parts of the whole, to molecules, to atoms, to particles, to quarks, at each stage we see a simpler model, each stage is less complex than the previous level. All of which is in perfect agreement with the Big Bang model that describes how all matter is built up from simple to more complex elements, stage by stage. When breaking down complex objects into smaller parts, it would come as a bit of a surprise if suddenly an entire universe popped up at even smaller scales than wave energy. Entire universes tend to be a bit complex!

If string theory is shown to be correct, then tiny loops, or strings, of vibrating wave energy may be the smallest, but they are not particles anyway, and strictly speaking quarks aren't either, as they cannot exist independently outside of a particle.

Stephen Hawking in his latest book "The universe in a Nutshell" (2001. page 176) describes the smallest possible size in our universe, termed the Planck length (after Max Planck, the famous German physicist) as being in the region of a millimetre divided by a hundred thousand billion billion billion. If we were able to probe to even smaller sizes, (which is not feasible as it would require particles that reside in black holes) we would not find anything smaller, we would, according to M-theory, see the other 6 or 7 dimensions of the 10 or 11 dimensions that go to make up our universe, only 4 of which are currently observed by us (length, breadth, height and time). These extra dimensions of our universe are curled up so small that we are unable to detect them, but the mathematics of theoretical physics have long said they must exist as part of the fabric of our universe. Please note that they are not 'other dimensions' with other universes, merely the smallest possible unit of our universe. I think we can quite safely rule out infinite smallness.

Likewise some claim that our universe is but an atom of another 'mega-universe', thus giving an infinite expansion in size. If our universe is finite (as indeed it is believed to be) then anything at all may be postulated as to what it may be that it exists in and is expanding into. You could for example propose that our universe is indeed but an atom of a 'mega-universe' but equally you could propose that it as an atom of a mega-donkey's hind leg, but as neither hypotheses is testable or falsifiable there is little point in proposing them. That's why this argument regarding infinity is restricted to just our known universe. However, although the idea that our universe may be a single atom within a 'mega-universe' may have some appeal, it is rather fanciful, and our universe bears absolutely no resemblance whatsoever to the atoms that we observe, or indeed to the particles within those atoms. Sorry, sounds fun but just doesn't match up with observation.

I do not believe that infinity exists in our universe, I think it only exists in mathematics. You may think differently of course, but I think you will agree that it would take some explaining!

Acknowledgements:
Steven Hawking *"A Brief History of Time"*, and *"The Universe in a Nutshell"*, Ivars Peterson *"The Mathematical Tourist"*

Throughout Chapter 1, the 'Science' chapter, I have included an 'Acknowledgements', as above. The purpose of this is twofold. It indicates the main source of my information from books, and serves as a guide for further reading should the reader be interested. There is also a Bibliography at the end of the book, but this method avoids the annoying and intrusive system of footnotes and references. However, It may seem as if all my information has been derived from the few books I mention, but these are simply the ones that happen to be in my personal collection, the others I cannot remember.

Keith Mayes.

02. Where did the universe come from?

"He took the golden Compasses, prepar'd In Gods Eternal store, to circumscribe This universe, and all created things: One foot he center'd, and the other turn'd Round through the vast profunditie obscure, And said, thus farr extend, thus farr thy bounds, This be thy just Circumference, OWorld."
Milton, *Paradise Lost,* Book VII

The Big Bang theory is an attempt to describe the creation and evolution of the universe. The theory appears to match observations, and the theoretical physics appear to hold back through time to within a tiny fraction of a second after the creation of the Big Bang. Beyond that the theory cannot explain how the Big Bang singularity came into existence. (See "Is the Big Bang Theory Correct?") Indeed, it is really pointless to attempt to go back beyond the Big Bang, it is meaningless to ask what came 'before' because there is no 'before'. Time itself came into existence with the Big Bang.

Stephen Hawking in 'A Brief History of Time' describes how time and energy came into existence with the creation of the Big Bang, but that the laws of science break down at the singularity preventing us from looking further back in time. As far as the Big Bang theory is concerned it is meaningless to look back beyond the Big Bang, nothing existed. Perhaps so, but it's very unsatisfactory to have a theory of the creation and evolution of the universe that does not explain where the universe, the Big Bang, actually came from, it must have come from something. Or must it?

Assuming that prior to the Big Bang there was absolutely nothing I will start from there, if on the other hand there was something, it would be necessary to explain where that 'something' came from. So in order to try and explain where the universe came from I will start with an attempt to describe a model of 'Nothing' that could contain the universe. If that succeeds I will then look at the problem of how the universe could be created from 'Nothing'. It should prove interesting to see where it leads us, or if we will have to be content with the idea of the universe starting from the Big Bang.

Before I begin however, I think we should agree on a few simple ground rules to try and keep the argument logical. Without the constraints of logic we could simply conjure up any description or semi-mystical event we wish in our attempt to make our model work, which would render the argument rather pointless. So here are the rules:
1) Once a definition has been made it can not be changed without starting a new definition.
2) An event must be possible within the framework of known science.
3) All events must follow a logical order within the given definition.

I will divide this model of 'Nothing' into two sections. Firstly I will attempt to create a 'working model' of 'Nothing', and if that appears to be successful then attempt to include the universe within it.

Definition of 'Nothing'.

The use of the word 'Nothing' has a very special meaning in this context, unlike our every day use of the word. It means here quite literally nothing, the complete absence of everything. By definition then 'Nothing' must be an infinite void. If 'Nothing' exists it would have to be infinite. This is a result of it not being allowed any boundaries, as a boundary would place a limit on 'Nothing's' size and furthermore would also indicate that there was something existing on the 'other' side of the boundary, apart from the boundary itself existing. This also, it should be noted, excludes anything existing in any other dimension, or dimensions, as a dimension would be a boundary. 'Nothing' then, excludes all possibility of anything else existing, anywhere.

I hope I have made this point absolutely clear, this is what having 'Nothing' would mean, absolutely nothing

15

anywhere, an infinite void. The only conclusion I can draw from that is 'Nothing' cannot exist, because we do.

Could 'Nothing' have existed in the past? No, because if it had existed in the past, then some event must have taken place to end it. An event would be impossible in 'Nothing', so 'Nothing' could never have existed because we do, and as our universe now exists, 'Nothing' can never exist in the future either. Why could an event not happen in 'Nothing'? Because apart from the obvious that there is nothing to happen, an event would create and require a moment in time. There can be no time in 'Nothing' as relativity describes time as just another dimension. See "What is Time?"

As for Time, without it 'Nothing' must have always existed, it cannot have a beginning or end because either would create a moment in time. It would in reality be meaningless to ask how long 'Nothing' has existed and how long it will continue to exist, it would be eternal and unchanging. Again, because we exist, 'Nothing' could not have had an existence because the creation of the universe would have required a significant change, thus contravening an unchanging 'Nothing'. We will look at this idea of creation in more detail later.

'Nothing' cannot have any laws of physics because there is nothing to apply those laws to, also the very concept of having laws contravenes our description of 'Nothing'. In the absence of any basic laws, let alone matter, how could anything be created? Once again, because we exist 'Nothing' could not have.

Could the universe have been created in 'Nothing'? No, for the reasons stated above. However, just for the sake of argument, let us imagine it was. If the universe was created in 'Nothing' then where was it 'put'? If somewhere 'outside' of 'Nothing', this would require an 'outside' to pre-exist, but it could not because that would require a boundary. It cannot be 'put' within 'Nothing', because containing a universe would no longer be within our definition of 'Nothing'.

So far we have discovered that by using the simple definition of 'Nothing' as being an infinite void we have placed the following conditions on it:

1) It must be timeless.
2) It must have always existed and could not have been created.
3) It is unchanging.
4) Nothing else can exist.
5) It is unable to create anything.

We have now concluded that 'Nothing', when described as an infinite void, could never have existed because we do. There is however nothing wrong with the definition itself, the existence of 'Nothing' as an infinite void would appear to be logical, more than that, it has to be that way, 'Nothing' could not have any restraints of size or time placed upon it.

We now need to change our definition of 'Nothing' in order that it may contain the universe.

A new definition.

We will retain the description of 'Nothing' that we had before, as an infinite void, keeping it exactly as it was, except for one change. We will now allow it to contain the universe.

Our definition of nothing will now read: "'Nothing' is an infinite void, nothing else can exist except for the universe that is contained within it."

We can now think of the universe as a tiny (or huge as you like, there is nothing to compare it with) 'bubble' existing in an infinite 'Nothing' and expanding into it. This model rather conveniently does away with the need to have a moment of creation for the universe because within 'Nothing' time does not exist. Without time it would be meaningless to ask when the universe was created, it was simply there all the time, existing in the same way as 'Nothing', as it always has. Within the universe of course time does exist, as does everything else. With this description of 'Nothing' its existence, and that of the universe, is now possible. Or is it?

What does it mean to say the universe was always there? We believe it started with the Big Bang, but can we say the Big Bang was always there? This doesn't seem logical to me, it needed to have actually come into existence at some point. Let's step back a little and look at the creation of the Big Bang from the viewpoint of a 'perfect observer' in 'Nothing'. At the moment of creation what would our 'perfect observer' see? Nothing at all! The universe is self contained, nothing at all can escape from it into our 'Nothing', our observer would notice no change whatsoever! As no detectable change at all has occurred from the viewpoint of 'Nothing', and no change could ever be detected regarding the expanding universe, no 'real' change has occurred, It may help here to visualise the Big Bang as an infinitely small event in the unimaginable vastness of an infinite void. In other words, a singularity, as indeed it is believed to have been. Therefore our definition of an unchanging timeless 'Nothing' is still valid. A read of Stephen Hawking's "A Brief History of Time" will clarify my point about nothing escaping from the universe and the Big Bang starting as a singularity.

Okay, I admit that I am on somewhat thin ice here suggesting that the creation event within 'Nothing' would not contravene our definition of an unchanging timeless 'Nothing' simply because 'Nothing' could not detect it. I will come back to the problem of how the Big Bang started later, at this point I am merely attempting to include the universe within 'Nothing'.

Let's now look at the implications of an infinite 'Nothing' containing an expanding universe, ignoring for now the actual creation. We will consider two possible problems, expansion and infinity.

1) Expansion. Can the universe be described as expanding? From our viewpoint within the universe, yes. From our 'perfect observer's' viewpoint in 'Nothing', no. Why not? Because, a) As stated above our observer can have no knowledge of the universe, and b) What is it expanding in relation to? 'Nothing' does not contain anything, other than the universe, so

there is no possible way to determine either the size, or the expansion of the universe, as both can only be measured in relation to something else. Size or expansion are meaningless terms here. This would appear to suggest that from within the universe things are as they appear to be, but from the point of view of our perfect observer in 'Nothing', the universe does not exist! Furthermore, with the absence of time in 'Nothing' the fact that it contains an ageing expanding universe is meaningless from the perspective of 'Nothing'. So far so good, our 'Nothing' is still intact; from the point of view of our infinite 'Nothing' it still contains nothing! (The creation event, if it actually happened, still needs explaining however)

2) Infinity. We now have a picture of 'Nothing' as being an infinite void, containing an expanding universe that it has no knowledge of, but is it still infinite? We have not put any restrictions on 'Nothing's' size, it is still infinite, but it contains a universe, so surely that puts restrictions on its 'completeness' - 'Nothing' is 'barred' from the area containing the universe. I think we are still okay here, to contain the universe is within our definition, but as to whether or not we have somehow a little less infinity is open to question, but it does not contradict our definition. I can see no reason why an infinite 'Nothing' cannot contain a finite universe. For a fuller argument on Infinity see "Does infinity exist?"

How is our new definition of 'Nothing' holding up? 'An infinite void, nothing else can exist except for the universe that is contained within it'. I would suggest that so far its holding up pretty well. I have not been able to overturn it on the grounds of logical argument. It could exist providing that the Big Bang took place within it. However, there is still a major hurdle to overcome, what caused the Big Bang and how could it form out of nothing? Without introducing a mysterious source of energy into the equation, as a magician might pull a rabbit out of a hat, it simply cannot be done, it's as simple as that. It's logically and scientifically

Keith Mayes

impossible to produce something from nothing. I realise that in Quantum Mechanics it is (arguably) possible, but that is in an already existing universe, not in 'Nothing'. Having said it's impossible we are left with a paradox, it has happened, we ARE here. There are only three logical conclusion to be drawn from this, assuming of course that our definition of 'Nothing' is valid.

1) The universe did not come from 'Nothing', it came from something. Taking this route however offers no explanation either, we would still need to explain where this new 'something' came from. We will therefore apply Ockham's razor and cut it out of our reckoning because it only adds to the complexity of the argument without adding any benefit, we gain nothing at all by introducing it. We may as well try to resolve the problem of the Big Bang coming from 'Nothing' rather than push it back a few steps and then try to solve it. We will therefore discard this idea.
2) We have to introduce a mysterious source of energy. I am forced to employ this highly undesirable tactic to make the creation of the universe possible. No matter how much I dislike the idea of using it I must, we do exist, so the universe must have been created out of nothing.
3) The universe did not have a creation event, it always existed.

So what is this mysterious source of energy that we are compelled to introduce? Many people will say that it is God and that He always existed. We either accept that or accept that the universe itself must have always existed.

We are now left with just these two possible solutions, either God created the universe and He always existed, or the universe itself always existed. The solution requires that something has always existed in order to avoid the problem of creating something out of nothing. The choice of introducing God is purely a matter of faith, for if we accept that God could have always existed then why not the universe? From a logical point of view within this model we do not need the existence of God, God is just a further complication that in turn would require to be created. If we ruthlessly apply Ockham's razor to the idea of introducing God into the model we are left with the universe always existing. However, for those of you of a religious nature allow me to make myself clear. I am NOT saying that God does not exist, only that the idea of introducing God into the equation is not necessary in order to make it work. See "Are all religions false"?

I know that some would argue that God is necessary as a Creator and Grand Designer of the universe, but I disagree. The universe can simply be the way it is by pure chance alone, it need not have been designed to be the way it is. For those that argue that the universe requires such a high degree of 'fine tuning' for things to be so well suited for our own creation and evolution that it could not have happened by chance alone, I disagree again. If the universe were not so well suited for us then we wouldn't be here to argue the point! The fact that we are here does not mean that the entire universe was designed just for our benefit. See "Is there a reason for our existence"?

All of the above would seem to suggest that the universe has always existed. I appreciate that the idea seems unsatisfactory to our way of thinking, but our way of thinking is probably part of the problem. In our universe we take for granted cause and effect, in that order. Everything we know of happens that way and even our minds work that way! Our very existence would not be possible if it were the other way round. When therefore we try to contemplate the idea of something always existing we simply cannot manage to understand it, we are seeking a 'cause' for the 'effect' of the universe existing. The universe however is different to us, it exists in 'Nothing', whereas we of course exist in the universe. There is no cause and effect in a timeless eternal infinite Nothing.

According to our definition of 'Nothing' as being timeless, then in order to contain the universe, the universe MUST have always existed within it. It is not possible for it to have been CREATED within it for that would require a moment in time. It is not a matter of convenience to suggest this idea, it is the way it simply has to be.

If however you are uncomfortable with the concept of anything having always existed then I see no solution at all, because you will simply have to accept that at some point something came from nothing, and personally I find that prospect totally unacceptable. Either that or you have to conclude that the universe does not exist! And that could be right.

Within the description of the Big Bang there are three main cosmological models. The 'open' universe that will expand forever, the 'flat' model that will come to a halt, or the 'closed' model that will re-collapse, possibly 'bouncing' back into another cycle of expansion. If the universe is closed it is possible that it will 'bounce' back cycle after cycle, forever. This idea of an eternal universe expanding and collapsing and re-expanding for ever is my preferred choice, but purely on aesthetic grounds. I realise of course that the arguments are still bouncing back and forth (sorry) as to which cosmological model is correct.

So after all the arguments I have made, what model do I prefer to describe where the universe came from? I prefer 'An infinite, eternal, unchanging 'Nothing' that has always existed and has always contained a finite but unbounded closed universe that constantly changes but is itself eternal'. In this model the Big Bang is NOT required as a creation event, it is merely a phase in the cycle of an eternally expanding and collapsing universe and has no special significance. There is no need to look beyond the Big Bang, there is only a previous cycle beyond it, and no need to say it is meaningless to try to look beyond it.

'Nothing' would appear to the casual observer to be a 'natural' state, but as I have outlined above, it seems to me it is not, it would appear to be a very special state.

With the model for 'Nothing' that I have described, it would appear to be possible to exist and to contain the universe, but it still does not give an explanation of how the universe could be created from nothing. This problem appears to be insurmountable. I can not 'fix' my theory to explain such an event and it would seem to suggest that the universe did not come from nothing but must have always existed or never existed! I tend to favour the view that the universe does exists, but of course we have no proof that it does!

It may be possible that we have not grasped the concept of 'Nothing'. Perhaps to exist it requires a structure, its own form of 'space', but I am not going to go down that particular road because that is not the 'Nothing' that I began with as a model of how it may exist. That would be a entirely new theory.

I honestly think that trying to explain where the universe came from is something we will never be able to do, we are contained within the universe and our understanding is restricted to the universe, anything else is guess work. That aside, this is the best attempt of describing where the universe came from that I could come up with: An infinite eternal unchanging nothing that has always existed and has always contained a finite but unbounded closed universe that constantly changes but is itself eternal. (Doesn't exactly roll of the tongue does it!)

With this model I can detect only one possible problem (I may be wrong of course, you may find many!) and that is the acceptance of the universe having always existed. If I could present a theory that proved this I would expect a Nobel Prize at the very least! Having said that I would suggest that the route to take in order to establish the concept of 'always' requires a more precise understanding of exactly what time is.

In the meantime, to answer the original question 'Where did the universe come from?' I believe that it didn't come from anything, it always existed. To say that I am unhappy with this concept is an understatement, but I am stuck with it because at this time I am unable to think of a viable alternative.

Of course my suggestion is just a model, created for the purpose of argument and discussion only, and I do not pretend for one minute that it is anything like the real thing, that, I am sure, will be much more surprising. It may be that it all exists only in our minds!

If you do not like my version, and why should you, why not try and come up with a better working model? You have a completely free hand!

Acknowledgements:

Stephen Hawking *"A Brief History of Time"*, and "*The Universe in a Nutshell"*, John Barrow *"The Origin of the Universe"*, James Cornell *"'Bubbles, voids and bumps in time: the new cosmology"*, Timothy Ferris *"The Whole Shebang"* Brian Greene *"The elegant universe"*, Stephen Hawking and Roger Penrose *"The nature of space and time"*, Michael Riordan and David Schramm *"The shadows of creation"*.

03. Is the Big Bang theory correct?

"The evolution of the world can be compared to a display of fireworks that has just ended; some few red wisps, ashes and smoke. Standing on a cooled cinder, we see the slow fading of the suns, and we try to recall the vanishing brilliance of the origin of the worlds."
Lemaitre

An overwhelming weight of evidence has convinced cosmologists that the universe came into existence at a definite moment in time, some 13 billion years ago, in the form of a superhot, superdense fireball of energetic radiation. This is known as the Big Bang theory. Until the arrival of the Big Bang theory the universe was believed to be essentially eternal and unchanging, represented by the Steady State model. The first clear hint that the universe might change as time passes came in 1917 when Albert Einstein developed his General Theory of Relativity. Einstein realised that his equations said that the universe must be either expanding or contracting, but it could not be standing still, because if it were then gravity would attract all the galaxies towards one another. This was, at the time, a revolutionary concept, so revolutionary that Einstein refused to believe it and introduced his infamous 'cosmological constant' into the equations so that the sums agreed that the universe could be static. He later claimed it was the biggest blunder of his career. It was in 1920 that Edwin Hubble discovered that the universe was expanding by measuring the light from distant galaxies. This discovery was followed in 1927 by Georges Lemaitre, a Belgian astronomer, who was the first person to produce a version of what is now known as the Big Bang model.
It is necessary to understand that the Big Bang did not begin as a huge explosion within the universe, the Big Bang *created* the universe. A popular misconception is that it happened within the universe and that it is expanding through it. This causes people to wonder where in the

universe it started, as if by running the clock backwards we would reach the point where all the galaxies come together in the centre of the universe. The universe does not have a centre, any more than the surface of a sphere has a centre, there is no preferred place that could be termed the centre. I know this sounds odd, it must have a centre, mustn't it? The problem we have here is we are trying to visualise the universe in the standard 3 dimensions that we are familiar with and therefore expect to find a centre to an expanding sphere. The universe, however, is not an expanding 3 dimensional sphere, it contains also the dimension of time (see 'What is Time?') and many other dimensions as well. By way of an illustration imagine a balloon with dots painted on the surface to represent the galaxies. If the balloon is now inflated we can see that all the dots are moving away from one another, just as the galaxies are in the real universe, and we can also see that on the surface of the balloon there is no centre point from which all the galaxies are moving away from. I am not suggesting that we are existing on the 'outside' of an expanding bubble, only that we cannot visualise the entire expanding universe.

Let's begin with a brief look at how the Big Bang describes the creation and evolution of the universe before moving on to some of the evidence to support the theory and the problems associated with the theory.

The Big Bang theory

The standard model of the Big Bang theory proposes that the universe emerged from a singularity, at time zero, and describes all that has happened since 0.0001 (10^{-4}) of a second after this moment of creation. The temperature of the universe at that time was 1,000 billion degrees Kelvin (10^{12}) and had a density that of nuclear matter, 10^{14} grams per cubic centimetre (the density of water is 1 gram per cubic centimetre). Under these extreme conditions, the photons of the 'background' radiation carry so much energy that they are interchangeable with particles. Photons create pairs of particles and antiparticles which annihilate one another to make energetic photons in a constant interchange of energy in line with Einstein's equation $E = mc^2$. Because of a small asymmetry in the way the fundamental interactions work, slightly more particles were produced than antiparticles - about one in a billion more particles than antiparticles.

When the universe had cooled to the point that photons no longer had the energy required to make particles, all the paired particles and antiparticles annihilated, and the one in a billion particles left over settled down to become stable matter.

One-hundredth of a second after time zero the temperature had fallen 90% to 100 billion K. By one-tenth of a second after time zero the temperature was down to 30 billion K. The temperature after 13.8 seconds was down to 3 billion K, and by three minutes and two seconds had cooled to 1 billion K, only 70 times hotter than the centre of the Sun today. At this temperature nuclei of deuterium and helium could be formed and stick together despite collisions with other particles.

During the fourth minute after time zero reactions took place that locked up the remaining neutrons in helium nuclei, as described by Gammow *et al* in 1940 and Fred Hoyle and others in the 1960's. This epoch ended with just under 25% of the nuclear material converted into helium, and the rest left behind as lone protons - hydrogen nuclei.

By just over 30 minutes after time zero, all of the positrons had annihilated with almost all of the electrons - with again one in a billion left over - to produce the background radiation proper, and the temperature had dropped to 300 million K, and the density was only 10% of that of water. At this temperature stable atoms were still not able to form.

The interactions between electrons and photons continued for 300,000 years, until the universe had cooled to 6000 K, roughly the temperature of the surface of the Sun, and the photons were becoming too weak to knock electrons off atoms.

Over the next 500,000 years the background radiation decoupled, and had no more significant interaction with matter. The Big Bang was in effect over, and the universe left to expand and cool. About 1 million years after time zero, stars and galaxies could begin to form. Nucleosynthesis inside stars convert hydrogen and helium to make heavier elements, eventually giving rise to our Sun, the Earth and ourselves.

This is only a very brief overview of the main points describing the evolution of the universe, a number of books have been published that describe just the first four minutes or less!

So how does it all stack up? How much evidence do we actually have to support the Big Bang model of the universe?

In support of the theory.

Einstein's Theory of Relativity. This is a theory of spacetime, offering a complete mathematical description of the universe. Relativity, along with Quantum Mechanics, (see "What is Quantum Mechanics?") is considered to be the most complete and accurate theory ever devised, mathematically describing such diverse phenomenon as the constant speed of light and the formation of black holes. Einstein's equations tell us - apart from many other things - that the universe is expanding, and that by going back in time there must have been a time when all the galaxies were very close together. And further back when all the stars must have been touching each another, merging to make one great fireball as hot as the inside of a star at 15 million degrees Kelvin (Kelvin is absolute zero temperature). Einstein's equations actually go further back than that, to a time when all the matter and energy of the universe emerged from a single point of zero size, a singularity. This is how the Big Bang theory describes the birth of the universe.

Expansion of the universe. One of the reasons the universe is believed to be expanding is because of the phenomenon known as 'red shift'. Light, or other electromagnetic radiation from an astronomical object may be stretched, (due to a number of reasons) making its wavelength longer. Because red light has a longer wavelength than blue light, the effect of this stretching on features in the optical spectrum is to move them towards the red end of the spectrum. If then the optical spectrum of a distant galaxy shows features that are shifted towards the red end of the spectrum (red shifted), it can be due to one or more of the following three reasons:

1) Motion. The galaxy is moving away from us, this is known as the Doppler effect. The same effect can be detected in sound. When a police car is speeding towards us the sound waves made by its siren are 'squashed' and the pitch sounds higher. As it passes us and starts to move away the sound waves are 'stretched' and the pitch sounds lower. In the 1920's Edwin Hubble observed that all galaxies (apart from a few local ones attracted towards our own and showing blueshift) show red shift. This indicates that the galaxies are all flying away from us, as in a Big Bang explosion.

2) Expansion of the universe. Einstein's famous equations show that the universe should be expanding, not because the galaxies were moving through space, but because the 'empty' space between them (spacetime) is expanding. This cosmological redshift results because the light from the distant galaxies is stretched by the amount that space expands while the light is en route to us. This also reveals that the Earth is not at the centre of the universe with all the galaxies moving away from us, but that due to the expansion of the universe, all the galaxies are moving away from each other, like painted dots on a balloon moving apart as it is inflated.

3) Gravity. This is also explained by Einstein's general theory. Light moving outwards from a star is moving 'uphill' in the star's gravitational field, and loses energy as a result. Because light cannot slow down - it always travels at the same speed - when it loses energy its

wavelength increases, in other words, it is redshifted. It does however, require a very powerful gravitational field for this effect to be measurable, such as created by a white dwarf star.

All three kinds of redshift can be at work at the same time. If we had telescopes sensitive enough to see light from a white dwarf star in a distant galaxy, the overall redshift in that light would be due to a combination of Doppler, cosmological and gravitational redshifts.

The fact that we can measure redshift in the light from distant galaxies tells us that the galaxies are receding from us, and from each other. It only takes a little logical deduction to conclude that as they are now all receding from one another, then at some finite point in the past (believed to be around 13 billion years or so) they must have all been at the same point.

Microwave Background Radiation. The universe is filled with a sea of radiation at a temperature of just over 2.7 degrees Kelvin, detectable at microwave radio frequencies both by Earth based radio telescopes and by instruments onboard artificial satellites. This is interpreted as direct evidence of the Big Bang fireball in which the universe was born, being the remnant of the superhot radiation from the fireball that has cooled down as the universe expanded. The discovery of the background radiation is therefore the most important observation made in cosmology since the discovery by Edwin Hubble that the universe was expanding. The existence of the background radiation, and its temperature, was accurately predicted by the Big Bang theory. When it was later discovered, by chance as it happens, this was yet another confirmation of the theory.

Nucleosynthesis of the light elements.
As the universe expanded and cooled, so the process of matter building began that led to the formation of stars, planets, galaxies etc. The process began with the simplest element, hydrogen, then helium, and then eventually onto

more complex elements. The observed abundance of hydrogen, the simplest element and the most common in the universe, followed by helium, is yet further confirmation of the Big Bang theory.

The study of stars reveals how their internal nuclear interactions cause simple atoms, such as hydrogen and helium, to create more complex elements. Stars are in fact gigantic matter producing factories, converting hydrogen and helium into carbon and heavier elements by the process of nucleosynthesis deep within their interiors. If stars did not exist, we would not be here, our own atoms that make us, were formed by the stars, and by the supernovas at the death of certain size stars. All as described by the Big Bang theory.

Formation of galaxies and large-scale structure

the Big Bang model provides a framework in which to understand the collapse of matter to form galaxies and other large-scale structures observed in the universe today. At about 10,000 years after the Big Bang, the temperature had fallen to such an extent that the average density of the universe began to be dominated by massive particles, rather than by light and other radiation. This change meant that the gravitational forces between the particles could begin to take effect, so that any small perturbations in their density would grow. These small perturbations led to the formation of galaxies.

These are the principle observed phenomenon that go to support the Big Bang theory. Is it proof enough? Do we have a universe that was born out of a singularity, that is expanding and cooling, and is therefore finite in both age and size? See "Does infinity exist?" for an explanation of a finite universe.

The Big Bang theory clearly still has a long way to go in order to be able to explain the origin of the universe. That is not to say that the theory as it stands is in any way wrong. The staggering amount of evidence in support of the Big

Bang theory is simply overwhelming. So much so that the theory cannot now be overturned. What is known to agree with the theory today cannot be changed tomorrow, by any theory, to make it disagree.

The situation is such that any new theory, far from displacing the Big Bang, would have to incorporate it. In other words, it can only be improved upon in much the same way that Einstein incorporated Newton's theory of gravity into his own theory of relativity. Relativity did not overthrow Newton's theory, it incorporated and developed it.

What problems does the Big Bang theory have?

Right now, the only serious problem facing Big Bang cosmology is that the NASA COBE satellite has shown that the cosmic background radiation is slightly lumpy in a way predicted by a version of Big Bang cosmology called 'inflationary Big Bang cosmology'. This version, however, requires that the universe has an average density that is exactly its 'critical density' given how fast it is currently expanding.

But when astronomers use a variety of independent observational techniques to measure what the density of our universe is, the numbers seem to come up short by a factor between 2 and 5. The recent study of supernovae located some 5 billion light years away have, again, indicated that the universe seems to have about five times less density than inflationary cosmology demands that it must have to be consistent with the COBE measurements.

Something is not adding up, but too many other things DO add up to favour big bang cosmology that no astronomer is even remotely suggesting that Big Bang cosmology is incorrect; only that we still do not know exactly the values of the specific parameters that define Big Bang cosmology.

Recently there was a seeming problem with the age of the universe turning out to be shorter than the ages of the oldest stars, but the Hipparcos Satellite re-calibrated the distance estimates, and now star and cosmos ages are within about 10 percent of each other. The

COBE/supernova problem may be resolved once we get more distant supernova to study.

I realise that some people do not support the Big Bang theory. Okay, agreed, it is 'only a theory.' It is also 'only a theory' that the Earth orbits the Sun, and not the other way around. But eventually, the mass of accumulated evidence for 'only a theory' becomes so powerful and persuasive it becomes impossible to ignore, whether or not you happen to like it.

Acknowledgements:
John Gribbin *"Companion to the Cosmos"*, John Barrow *"The Origin of the universe"*, Timothy Ferris *"The Whole Shebang"*, Brian Greene *"The Elegant universe"*, Steven Weinberg *"The First Three Minutes"*, Stephen Hawking *"A Brief History of Time"*.

04. What is Quantum Theory?

"Something unknown is doing we don't know what."
Sir Arthur Eddington's comment on the Uncertainty Principle in quantum physics, 1927

Quantum theory is bizarre. In order to try and understand it we need to forget everything we know about cause and effect, reality, certainty, and much else besides. This is a different world, it has its own rules, rules of probability that make no sense in our everyday world. Richard Feynman, the greatest physicist of his generation, said of quantum theory

'It is impossible, absolutely impossible to explain it in any classical way'.

Quantum theory is much more than just bizarre, it is also without doubt the most amazing theory in existence. If after reading this section you are not totally amazed by it, then the fault is mine, for I will have failed to reveal to you its deep underlying significance. This theory is not just about experiments and equations, it reveals something extraordinary about our very understanding of what constitutes reality.

This is a very complex theory, and in order to fully do it justice it would require a separate book in its own right. However, in order to grasp the basic principles involved it will suffice to study just three key experiments. The three experiments are generally known as: the 'Double Slit Experiment', Schrödinger 's 'Cat-in-the-Box Experiment' and the 'EPR Paradox'.

We will start with the famous double slit experiment as it demonstrates beautifully the central mystery of quantum theory. Quantum theory however, needs some introduction before we get too involved in the experiment.

The standard explanation of what takes place at the quantum level is known as the Copenhagen Interpretation, because much of the pioneering work was carried out by the Danish physicist Niels Bohr, who worked in

Copenhagen. Quantum theory attempts to describe the behaviour of very small objects, generally speaking the size of atoms or smaller, in much the same way as relativity describes the laws of larger everyday objects. We find it necessary to have two sets of rules because particles do not behave in the same way as larger everyday objects, such as billiard balls. We can, for example, say precisely where a billiard ball is, what it is doing, and what it is about to do. The same cannot be said for particles. They are, quite literally, a law unto themselves, and why this should be so is a source of much debate. The classic experiment to illustrate this is the famous double slit experiment, originally devised to determine if light travels as waves or particles. Feynman said of it:

'Any other situation in quantum mechanics, it turns out, can always be explained by saying, *"You remember the case of the experiment with the two holes? It's the same thing."'*

The double slit experiment.

If light travels as particles you can imagine particles of light (photons) as bullets fired from a rifle. Imagine a brick wall with two holes in it, each the same size and large enough to fire bullets through, with a second wall behind where the bullets will strike. After firing a few rounds you would expect to see on the second wall two clusters of hits in line with the two holes. This is of course precisely what you get with bullets, so if we get the same result with photons we can say they are particles.

Now imagine that instead of particles, that light travels as a wave, we can replicate that with a water tank. As the wave spreads out from its source it would reach both holes at the same time and each hole would then act as a new source. Waves would then spread out again from each of the holes, exactly in step, or in phase, and as the waves moved forward, spreading as they go, they would eventually interfere with one another. Where both waves are lifting the water surface upward, we get a more pronounced crest;

where one wave is trying to create a crest and the other is trying to create a trough the two cancel out and the water level is undisturbed. The effects are called constructive and destructive interference.

If we carried out this procedure with light instead of water, and if light travels as waves, then the pattern on the second wall would appear as an interference pattern of alternate dark and light bands across the wall. Particles, on the other hand, would produce two separate areas of light (where the bullets would hit). This experiment has in fact been carried out many, many times, with the same results every time, and the results are nothing less than amazing.

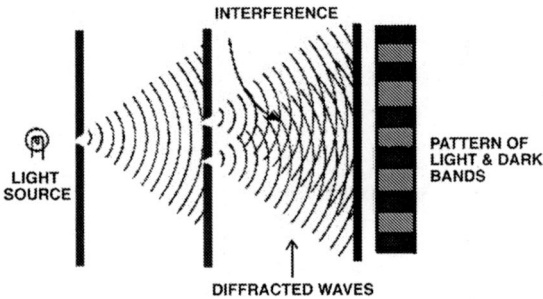

When the experiment is set up as shown in the above diagram, with both slits open, the resulting interference pattern clearly shows that light behaves as a wave. Now if that was all there was to it we could all fold up our tents and go home happy in the knowledge that light travels as a wave; but there is much more to it than that. This is where the word 'weird' can become over-used.

If the experiment is set up to fire *individual* photons, so that only one photon at a time goes through the set up, we would not expect the same interference pattern to build up; we would surely expect that a single photon would only go through one hole or another, it cannot go through both at the same time and create an interference pattern. So what happens?

If we wait until enough individual photons have passed through to build up a pattern - and this takes millions of photons - we do not get two clusters opposite the two holes, we get the same interference pattern! It is as if each individual photon 'knows' that both holes are open and gives that result. Each individual photon, passing through the set up will place itself on the wall in such a position that when enough have passed through they have collectively built up an interference pattern, when there cannot possibly be any interference!

If we repeat the experiment, this time with only one hole open, the individual photons behave themselves and all cluster round a point on the detector screen behind the open hole, just as you would expect. However, as soon as the second hole is opened they again immediately start to form an interference pattern. An individual photon passing through one of the holes is not only aware of the other hole, but also aware of whether or not it is open!

We could try peeking, to see which hole the photon goes through, and to see if it goes through both holes at once, or if half a photon goes through each hole. When the experiment is carried out, and detectors are placed at the holes to record the passage of electrons through each of the holes, the result is even more bizarre. Imagine an arrangement that records which hole a photon goes through but lets it pass on its way to the detector screen. Now the photons behave like normal, self respecting everyday particles. We always see a photon at one hole or the other, never both at once, and now the pattern that builds up on the detector screen is exactly equivalent to the pattern for bullets, with no trace of interference. As if that was not bad enough, it gets even worse! We do not need place detectors at both holes, we can get the same result by watching just one hole. If a photon passes through a hole that does not have a detector, it not only knows if the other hole is open or not, it knows if the other hole is being observed! If there is no detector at the other hole as well as the one it is passing through, it will produce an interference pattern, otherwise it will act as a particle. When we are watching the holes we can't catch out the photon going through both at once, it will only go through one. When we are not watching it will go through both at the same time! There is no clearer example of the interaction of the observer with the experiment. When we try to look at the spread-out photon wave, it collapses into a definite particle, but when we are not looking it keeps its options open.

What the double slit experiment demonstrates is this: Each photon starts out as a single photon - a particle - and arrives at the detector as a particle, but appears to have gone through both holes at once, interfered with itself, and worked out just where to place itself on the detector to make its own small contribution to the overall interference pattern. This behaviour raises a number of significant problems! Does the photon go through both holes at the same time? How does a photon go through both holes at the same time? How does it know where to place itself on the detector to form part of the overall pattern? Why don't all the photons follow the same path and end up in the same place?

As a possible explanation it could perhaps be said that this is just one more example of the extraordinary nature of light, after all it does have some very unusual properties. Photons have no mass for example, a very odd property! Light is also unique in that it always travels at the same speed. However you move, and however the light source moves, when you measure the speed of light you always come up with the same answer. By way of comparison, two cars approaching each other and each having a speed of 30 mph will be approaching each other at a speed of 60 mph. Two light beams, both travelling of course at the speed of light, will be approaching each other at the speed of light, not twice the speed of light. Perhaps the weird behaviour of photons in the experiment is due to the weird nature of light. Unfortunately further experiments have demonstrated that this is not the case. Electrons have been used instead of photons, and they not only have mass, they have an electric charge, and furthermore they move at different speeds depending on circumstances, like normal everyday objects.

The double slit experiments still gives the same result using electrons as it does using photons; electrons also alter their behaviour depending on whether or not they are being observed. The experiment has even been performed using atoms, again with the same result, and atoms are large enough to be individually photographed, they are very real solid objects. This odd behaviour of particles is a very *real* phenomenon.

The double slit experiment is not simply an oddball theory that has no application in the real world. This strange behaviour of particles lies at the very heart of our understanding of the physical properties of the world. Quantum theory is used in many applications, including television and computers, and even explains the nuclear processes taking place inside stars.

One possible explanation for quantum weirdness is a theory concerning the nature of the wave that is passing through the experiment. The key concept of the theory, which forms a central part of the Copenhagen Interpretation, is known as the 'collapse of the wave function'. The theory seeks to explain how an entity such as a photon or an electron, could 'travel as a wave but arrive as a particle'. According to the theory, what is passing through the experiment is not a material wave at all, but is a 'probability wave'. In other words, the particle does not have a definite location, but has a probability of being here or there, or somewhere else entirely. Some locations will be more probable than others, such as the light areas in the interference pattern for example, and some will be less probable, such as in the dark areas. In this theory, an electron that is not being observed does not exist as a particle at all, but has a wave-like property covering the areas of probability where it could be found. Once the electron is observed, the wave function collapses and the electron becomes a particle. This theory rather neatly explains the behaviour of the particles in the double slit experiment. When we are not looking at the particle, the probability wave, of even a single particle, is spread out and will pass through both slits at the same time and arrive at the detector as a wave showing an interference pattern. When we observe the electron by placing detectors at the slits, it is forced into revealing its location which causes the probability wave to collapse into a particle. If the theory is correct, its implications are staggering. What it suggests is that nothing is real until it has been observed!

Nothing is real until it has been observed! This clearly needs thinking about. Are we really saying that in the 'real' world - outside of the laboratory - that until a thing has been observed it doesn't exist? This is precisely what the Copenhagen Interpretation is telling us about reality. This has caused some very well respected cosmologists (Stephen Hawking for one) to worry that this implies that there must actually be something 'outside' the universe to look at the universe as a whole and collapse its overall wave function. John Wheeler puts forward an argument that it is only the presence of conscious observers, in the form of ourselves, that has collapsed the wave function and made the universe exist. If we take this to be true, then the universe only exists because *we* are looking at it. As this is heading into very deep water I think we will have to leave it there and move on to the next experiment. Just keep it in mind though as we shall be return to it later on in the book.

Schrödinger 's 'Cat-in-the-Box Experiment'

According to the Copenhagen Interpretation, the probability wave of an electron requires the act of observation by a conscious observer to collapse it into a definite particle, and thus have a definite location. We can imagine a closed box containing just a single electron. Now until someone looks in the box, the probability wave associated with the electron will fill the box uniformly, thus giving an equal probability of finding the electron anywhere inside the box. If a partition is introduced into the middle of the box that divides it into two equal boxes, still without anyone looking inside, then common sense tells us that the electron must be in one side of the box or the other. But this is not the case according to the Copenhagen Interpretation; that says that the probability wave is still evenly distributed across both half-boxes. This means that there is still a 50:50 chance of finding the electron in *either* side of the box. When somebody looks into the box the wave will then collapse and the electron will be noticed in one half of the box or the other, but it will only at the moment of observation 'decide' which half it will be in. At the same time the probability wave in the other half of the box vanishes. If the box is then closed up again, and the electron no longer observed, its probability wave will again spread out to fill the half box, but cannot spread back into the other half of the box that was empty.

The way that a quantum wave moves is described by Erwin Schrödinger's wave equation and describes the probability for finding a photon, or electron, at a particular place. Schrödinger did not however, go along with the 'collapse of the wave function' theory, he thought it was a nonsense, and designed 'thought experiments' to prove his point. In an attempt to demonstrate the foolishness - as he saw it - of quantum theory, Schrödinger devised the cat-in-a-box thought experiment.

In Schrödinger 's original thought experiment he used radioactive decay because that also obeys the rules of probability. We however, shall use our box with the partition and electron again, as we are now familiar with it.

Imagine we have our box with the partition in place, and the electron's probability wave evenly spread between both halves of the box. We have now added a device that will, at a given time, automatically open up one half of the box to the room. There is a 50:50 chance that when opened the box will contain the electron that is now free to enter the room. The room is sealed and has no windows that would allow any outside observations to be made. Inside the sealed room there is a cat, a container of poisonous gas, and an electron detector. The experiment is so designed that if the electron detector detects an electron it will release the poisonous gas into the room, which would prove very unfortunate for the poor cat. If, on the other hand, that half of the box does not contain the electron, the poisonous gas will not be released into the room and our cat, henceforth

Keith Mayes

known as Lucky, will continue to enjoy good health, providing it keeps away from busy roads.

Taking a common sense view of the situation, we would say that when the experiment has run its course, and an observer enters the room, they will find the cat either dead or alive. But we already know enough about quantum theory to realise that common sense doesn't apply here, and instead we have to turn to the Copenhagen Interpretation for an explanation.

According to the Copenhagen Interpretation, when the lid of one half of the box is opened, it is not an electron, or not as the case may be, that is released into the room, but the probability wave of the electron as it has not yet been observed. This raises the question of whether or not the cat can be regarded as a conscious observer. If it can be then where do we draw the line? Would a fly or an ant count? How about a bacterium? As this is again getting into rather deep water, we will skip over this problem and continue with our experiment, otherwise we run the risk of becoming seriously side-tracked. So the probability wave spreads into the room, not an electron (or no electron). The electron detector is itself composed of microscopic entities of the quantum world (atoms, particles and so on) and the interaction of the electron with it would take place at this level, so the detector is also subject to the quantum rules of probability. Taking this view, the wave function of the whole system will not collapse until a conscious observer enters the room. At *that moment* the electron 'decides' whether it is inside the box or in the room, the detector 'decides' whether it has detected an electron or not, and the cat 'decides' whether it is dead or alive. Until that moment, according to the Copenhagen Interpretation, the cat is not either dead or alive, it describes the situation as a 'superposition of states'. Only the act of observation will cause it to become one or the other. Schrödinger described the situation as *'having in it the living and the dead cat mixed or smeared out in equal parts.'* The Copenhagen Interpretation does not allow for the room to actually contain a cat that is both dead and alive at the same time, or a cat that is neither dead nor alive, suspended in limbo. But contains *either* a dead cat or a live cat, *until someone looks*, and it is then that the actual reality of the situation is determined.

Cat lovers please note. This experiment has never been carried out, and never will be. This is not only because it would be a very cruel thing to do, but because it wouldn't prove anything. An observer upon entering the room would find either a dead cat or a living one, but could not observe what processes preceded this event. Any previous observation would of course defeat the object of the experiment.

The problems highlighted by the cat-in-a-box experiment raise some very deep questions. What for example are the requirements needed to qualify as a 'conscious observer'? Do the probability waves of particles spread out again when not observed and particles somehow become less 'real', as described by the Copenhagen Interpretation? Does the universe exist only because we are here to observe it? Could a cat really be in a 'superposition of states', either dead or alive until the moment of observation? This goes entirely against all our common sense experience of life, we would naturally conclude upon finding the cat alive that it had 'obviously' been alive all the time. Quantum theory is telling us that we could be very wrong in our thinking regarding what reality really is.

Quantum theory has yet another surprise in store for us, and this time it's not simply another bizarre phenomenon that challenges our common sense. This time it contradicts one of the central principles of Einstein's theory of relativity, that nothing can travel faster than the speed of light. As you can imagine, Einstein was not amused.

The EPR Paradox.

The experiment is so named because it was a thought experiment devised by Einstein, Boris Podolsky and Nathan Rosen. As with Schrödinger 's cat-in-the-box experiment, its purpose was to expose the 'foolishness' of the

Copenhagen Interpretation. The experiment focuses on the phenomenon of quantum theory known as 'non-locality', which concerns communication between particles. A pair of protons, for example, associated with one another in a configuration called the singlet state will always have a total angular momentum of zero, as they each have equal and opposite amounts of spin. Just as we have seen in the other experiments, the protons will not collapse their probability wave and 'decide' which spin to adopt, until they have been observed. If you measure the spin of one proton, according to quantum theory, the other proton instantly 'knows' and adopts the opposite spin. So far so good, we have come to expect this sort of behaviour from particles, so what is the problem with this particular experiment?

It is possible, and has been carried out in laboratory tests over a short distance, to split the particles apart and send them in opposite directions and then measure one of them for spin. The instant it is measured, and the spin determined, the other particle adopts the opposite spin. The time interval is zero, the event takes place instantaneously, even though the particles are separated, and theoretically would still do so even if they were separated by a distance measured in light years. This is what upset Einstein, the implication that particles could communicate at faster than light speed, it is impossible for this to happen according to Einstein's theory of relativity.

At the time this thought experiment was proposed, in the early 1930's, just about the time of Schrödinger 's cat-in-the-box thought experiment, it was not actually possible to physically carry out the experiment. Einstein did not live to see it turned into practical reality, which is probably just as well in light of the results produced. This experiment has now actually been carried out over a distance of 10 kilometres and confirmed as correct. Something here is taking place at faster than light speed, although exactly what seems to be a mater of some debate. Regrettably, due to its very nature, no meaningful communication could be made using such a device. Whether or not it will ever have any useful application remains to be seen, but that is not the point. The point is the experiment has proved Einstein wrong, faster than light speed, at least in the quantum world, is a reality. However, in classical physics - at sizes above that of atoms - relativity still remains unchallenged, nothing has been detected at faster than light speed.

As I said at the outset of this section, these three experiments highlight the basic principles involved in quantum theory. I also said they would amaze you, and I hope that you feel that I have kept my word. If you are not amazed by quantum theory, then blame me, for the theory is truly amazing and any disappointment you may have with it can only be due to my inability to do the theory justice.

One last thing you need to know about quantum theory, and that is Heisenberg's Uncertainty Principle. Heisenberg said that the electron was a particle, but a particle which yields only limited information. It is possible to specify where an electron is at a given moment, but we cannot then impose on it a specific speed and direction at the setting-off. Or conversely, if you fire it at a known speed in a certain direction, then you are unable to specify exactly what its starting-point is - or its end-point. The information that an electron carries is limited in its totally. That is, for instance, its speed *and* its position fit *together* in such a way that they are confined by the tolerance of the quantum. The following quote is from Dr. Jacob Bronowski's excellent book 'The Ascent of Man'. *"Heisenberg called this the Principle of Uncertainty. In one sense, it is a robust principle of the everyday. We know that we cannot ask the world to be exact. If an object (a familiar face, for example) had to be* exactly *the same before we recognised it, we would never recognise it from one day to the next. We recognise the object to be the same because it is much the same; it is never exactly like it was, it is tolerably like. In the act of recognition, a judgement is built in - an area of tolerance or uncertainty. So Heisenberg's principle says that no events, not even atomic ones, can be described with certainty, that is, with zero tolerance. What makes the principle profound is that Heisenberg specifies the tolerance that can be reached. The measuring rod is Max Planck's quantum. In the world of the atom, the area of uncertainty is always mapped out by the quantum."*

The principle of Uncertainty fixed once for all the realisation that all knowledge is limited, that there is no such thing as absolute certainty.

John Gribbin in "Companion to the Cosmos" says: *"It is this quantum uncertainty that allows electron-positron pairs (and other particle-antiparticle pairs) to appear out of nothing at all (out of the vacuum), provided they annihilate one another in the brief flicker of time allowed by quantum uncertainty. This is the source of Hawking radiation associated with black holes. It is even possible that the entire Universe was created in this way, by inflation out of a quantum fluctuation of the vacuum."*

What conclusions can we draw from these experiments?

We need to be very careful in drawing any conclusions from the results of these experiments. All we can say with any confidence is that if we set up the apparatus in a certain way it will produce a certain result. How we interpret those results, the meanings that we attach to them, is nothing more than our way of attempting to make sense of them, and need have no relationship at all to the actual reality of the situation. To imagine that a probability wave passes through both slits in the double slit experiment helps us to understand what may be happening, but it is in fact nothing more than proposing an idea that meets the criteria of what has been observed; there may be no such thing as a probability wave. It may be the case that we are completely missing some fundamental property of particles, a property that as yet remains undetected by our equipment and experiments. There may be things going on that we are completely unaware of.

What quantum mechanics tell us is that nothing is real and that we cannot say anything about what things are doing when we are not looking at them. In the world of quantum mechanics, the laws of physics that are familiar from the everyday world no longer work. Instead, events are governed by probabilities. Einstein was so disgusted by the whole notion that he made his famous remark, *"Quantum mechanics is very impressive. But an inner voice tells me that it is not yet the real thing. The theory produces a good deal but hardly brings us closer to the secrets of the Old One. I am at all events convinced that He does not play dice"*.

It seems illogical that we need two completely different laws to explain the behaviour of objects, dependent on how large or small the object is. Why is it that the laws of cause and effect, that work so well in the everyday world, breakdown in the world of the very small, when everything in the everyday world is made up of the very small?

It just does not make any sense, but like it or not, until a theoretical physicist comes up with a theory that incorporates both Quantum Mechanics and Relativity we just have to admit that we do not really know what is going on. However, one thing I am absolutely clear on is that an electron, or photon, doesn't 'know' anything, anymore than a frozen pea does. When you remove a frozen pea from the freezer and place it in a warm room you do not gasp in amazement when it defrosts and ask how did it 'know' to defrost. You do not try and trick it into not defrosting by leaving it in the freezer and turning off the freezer. This is of course because we understand the laws of thermodynamics. Particles do not 'know' anything!

When physicists ask the question, 'how does a particle 'know' something'? they are of course using the term loosely. What they are really asking is 'what are the forces acting upon the particle that we have not detected? What interactions are taking place that we have not detected?'

That is the problem. Something is going on at a level that we are completely unaware of. However, the idea of

47

probability waves as an explanation is nothing more than an attempt to describe what is observed in the quantum world by the Copenhagen Interpretation, and is of course a purely theoretical concept.

It may be possible that we need to develop a new form of logic to be able to describe what is happening at the quantum level. I will end this section with a quote from "*The world within the world*" by John D. Barrow, as it provides wonderful food for thought!

"*...the peculiar aspects of quantum reality...may be signalling to us that it is not just our physical theory that needs improving if we are to deal with the micro-world. Indeed, mathematics may not even be the language that most naturally describes what is happening. There has been investigations into the possibility that classic logic (where statements are either true or false) does not apply at the quantum level, but is replaced by a three-valued "quantum logic" which allows the additional status of "undecided" to be associated with a statement; thus a statement that is not true need not be false. In this way a different answer can be given to the question "what slit does the particle go through in the two slit experiment? However, the solution to the problems of quantum reality may be far more radical. It may require some other type of description: a new language that does for logic what logic does for mathematics.*"

Acknowledgements:
Much of the material here is condensed from John Gribbin's excellent books '*Schrödinger's kittens and the search for reality*', '*In search of Schrödinger's cat*'. '*Companion to the Cosmos*' and '*Blinded by the Light*'. Also '*Alice in Quantumland*' by Robert Gilmore, and '*QED The strange theory of light and matter*' by Richard Feynman. '*The Ascent of Man*' by J. Bronowski. "*The world treasury of physics, astronomy. and mathematics*" by Timothy Ferris

Any mistakes that may have been made in portraying quantum theory are mine.

05. What is Time?

"The most exciting phrase to hear in science, the one that heralds new discoveries, is not 'Eureka!' (I found it!) but 'That's funny ..."
Isaac Asimov

When we think of time we tend to think of the ways in which we measure the passing of time, such as a clock or watch, or perhaps a measured interval of time such as an hour or minute, but not of time itself. So what is time? Exactly what is it that we are measuring?

We can begin to answer the question with the basic description that we are measuring the interval between events, using units that we have chosen for the purpose. We may say, for example, that the next train will be due in 5 minutes. While this information may be very useful for telling us how late the train is when it eventually arrives, it does nothing to describe just what it is that we are measuring. We want to know exactly what the 'interval' is.
In order to investigate the nature of time it may help to break it down into four main questions.

1) How does time flow?
2) Does time flow in only one direction?
3) Is there a constant 'Universal' time?
4) Is time a 'real' dimension?

1) How does time flow?
We tend to perceive time as 'flowing', as though it were in smooth and perpetual continuous motion, but is this view correct? We have learned that at the quantum level energy is not released continuously - there is a limit to how small a change in energy an atom can experience - it is released in discrete quanta by the emission of a single photon. Could there be a limit to the change in time? This would mean that time would advance in small discrete steps and not

move continuously, in other words it would move in a similar way to watching the progress of a story on a film or video; the individual 'frames' of time may be so small that it only gives the appearance of being continuous. This can be tested experimentally by using sophisticated equipment to observe chemical changes taking place at very small fractions of a second. If time does move in small steps, then by probing ever smaller segments of time it may be possible to reach a limit at which these steps can be observed to take place.

Equipment has been constructed that can 'slice' moments in time small enough to capture a chemical reaction take place, rather like a 'freeze frame' picture. This requires an extremely small fraction of a second to observe the process taking place, and is called a Femtosecond. The method of observing such small moments of time is achieved using pulsating laser beams. How small is a Femtosecond? It is one thousandth of one trillionth of a second, and can be expressed as 1/1,000,000,000,000,000th of a second. To try and put this very small fraction of a second into perspective we can use a comparison. According to my calculations, a Femtosecond is to a second, as a second is to 32 million years! So dumbfounded was I by this comparison I checked it out three more times! Another comparison is to take the distance from the Earth to the Sun and slice it up into units of 0.15 of a millimetre, this being the same ratio as a Femtosecond is to a second. Even when dividing time up into this incredibly small unit there is no indication of time passing in discreet steps, it still appears to flow smoothly.

What conclusions may be drawn from these observations of time at the level of a Femtosecond? All we can say is that either time does flow smoothly and continuously, or if it moves in discrete steps we have not yet reached a level small enough to observe it.

In terms of pure research, scientists refer to an even smaller unit of time, called the Attosecond, which is one quintillionth (10^{-18}) of a second. This is a decimal point followed by 18 zeros and then 1 (0.0000000000000000001) and is a term used in photon research. The smallest measurement of time that can have any meaning within the framework of the laws of physics as understood today, is known as 'Planck Time', and is equal to 10^{-43} seconds. We can only describe the universe as coming into existence when it already had an age of 10^{-43} seconds. It may be that we still have some work to do in order to observe time moving from one 'frame' to the next, if that is indeed what happens.

The Hubble Space Telescope has been used to try and determine if time is continuous or not. Dr Richard Lieu and Dr Lloyd Hillman observed a number of galaxies at a distance of more than 4 billion light years from the Earth. They were looking for light patterns that shouldn't be present if the standard ideas from quantum theory apply to time. According to quantum theory the inherent uncertainty means that time (and hence speed) cannot be measured to infinite accuracy, but that it flows 'fuzzily' on the quantum scale. What they found was the images of the galaxies exhibited a sharp 'Airy' diffraction ring. This implies that the speed of light didn't change by more than 1 part in 10^{32} as it travelled to through space to reach us. If quantum theory of times are correct then it should not be possible to measure to this degree of accuracy. We may have to accept that time does flow smoothly and not in discreet steps.

Dr Lieu, February 2003. *"This discovery will present problems to several astrophysical and cosmological models, including the Big Bang theory of the universe. The Big Bang theory supposes that at the instant of creation, the quantum singularity that became the universe would need to have infinite density and temperature".* In light of these new findings it would appear that the conventional solution of arguing that the fuzziness of time smears out the singularity, keeping density finite, now seems impossible.

2) Does time flow in only one direction?
We perceive time as flowing from the past through the present and into the future. We have memory of past

events, but of course no memory of future events. Time provides us with a base line reference point in which events can be placed in order of occurrence, and in this manner we are able to establish that one event occurred before or after another, and this provides us with the so called 'arrow of time'. Interestingly, there is nothing in the laws of physics to suggest that time actually flows from the past through the present and into the future. What is it that gives time a definite direction, the arrow of time? To seek the answer we need to examine the laws of thermodynamics.

At a subatomic level there is no distinction between the past and the future. In a typical interaction involving subatomic particles, two particles may come together and interact in some way to produce two different particles, which then separate. According to the laws of physics there is no reason why these two new particles could not then interact and revert to their initial condition. By studying these particles it would be impossible to determine the order of events that took place, or indeed if any event had taken place. At this level there is no way to distinguish the past from the future simply by looking at each pair of particles.

In the macroscopic world - at the level detectable by our own senses - we are clearly able to discern the arrow of time. If we see a picture of a tumbler of water on a table, and another of a broken glass on the floor lying in a puddle of water, we know the order of events that took place. We know that broken tumblers never reassemble themselves and that spilled water will not gather together and place itself back in the glass. But why not? According to the known laws of physics every interaction involving the atoms of the tumbler as it smashes is reversible, as is the spilled water. But there is an inbuilt arrow of time, pointing from the past to the future, when we are dealing with complex systems which contain many particles. This distinction between past and future events can be expressed mathematically by the science of thermodynamics, which is based on analysis of the way things change as we 'move' from the past into the future.

The second law of thermodynamics, the most fundamental law of physics, states that the entropy of a closed system always increases, entropy being the measure of disorder. In other words, in a closed system - and the universe is a closed system - disorder will always increase, things will never arrange themselves to a degree of higher order. If for example you put together a jigsaw puzzle in a box to form the completed picture, then close the lid and shake the box, you would not expect to see the pieces rearrange themselves back to the complete picture again, no matter how long you shook the box. The explanation is simple statistics, there is only one possible correct solution, but many wrong ones, so we would expect to keep getting wrong ones. If we could keep trying for long enough then it is statistically possible that the puzzle may by chance put itself correctly back together again, but this is very unlikely, and the universe may not have enough time for this highly unlikely event to occur. Thus left to its own devices, a system will run to disorder, and not order, and this gives us an arrow of time.

Another arrow of time is the Big Bang, and this may be described as the ultimate arrow of time. No matter at what place or time you are in the universe, the Big Bang always lies in the past direction of time. We see the same arrow of time in the expansion of the universe. As the universe ages and expands galaxies are moving further apart. Where galaxies are closer together points in the past direction of time, and where they are further apart points to the future direction of time.

The first law of thermodynamics states that the amount of energy in a closed system cannot change. The total energy of the universe was determined at its creation, but what the second law tells us is that the total amount of 'useful' energy decreases. If and when all the stars and other sources of energy in the universe have given up their heat, the entire universe will be in a state of uniform temperature in which nothing ever changes. it will have suffered 'heat death'.

What have we learned from the study of thermodynamics in relation to the arrow of time? We have learned that the reason why events are reversible on the microscopic scale but irreversible on the macroscopic scale (why the arrow of time points only one way) is that the law of increasing entropy is a statistical law; a decrease in entropy is not so much forbidden as extraordinarily unlikely. Sounds similar to the quantum probability wave doesn't it? So the answer is that time does appear to flow in only one direction, on the macroscopic scale.

3) Is there a constant 'Universal' time?

If we were to take two atomic clocks and synchronise them to read the same time, we know that we could leave them to 'tick' away for a year and they would both still read the same time. But if we separated the two clocks and took one of them on an aeroplane journey around the world, then when compared to the other clock on return it would be seen to be running a fraction of a second slower. It will only be a very small fraction of a second, but the difference would be real, our globe-trotting clock will have run slower than our stay-at-home clock, travelling the world really does keep you younger! This is not just a theoretical concept, many different experiments, including the 'globe trotting clock', have been carried out and have proved the theory to be correct. So what's going on? The answer can be found in Einstein's theory of relativity, because that tells us that the faster an object moves the slower time runs, until at the speed of light time comes to a stop. This effect is known as 'time dilation', and is very small when travelling at day to day speeds, but becomes significant at relativistic speeds- speeds that are approaching the speed of light. For an example of time dilation at speeds that we are familiar with, an astronaut having spent a year aboard the space station will have aged 0.0085 seconds less than the rest of us that stayed on Earth - hardly eternal youth is it?

It is important to understand that relativistic speed, or any other speed come to that, does not effect the speed at which clocks run, it is time itself that slows down.

We have seen that the speed at which time passes varies in direct proportion to the speed of the observer, but there is another factor that needs to be accounted for, and that is gravity. Powerful gravitational fields, such as those found at the event horizon of black holes, also slow down time. Similarly, a clock at sea level will record time running slower than a clock at high altitude on a mountain.

To answer the original question, 'Is there a constant Universal time?', the answer is clearly no. We all experience time as passing at different speeds, relative to our speed in relation to one another and the strength of any gravitational field that we may happen to be in.

4) Is time a 'real' dimension?

The Big Bang theory describes how the universe was created from the Big Bang singularity, where all matter and space is contained in a single point of infinite density. At the moment of creation of the universe - the Big Bang - all matter, space and time came into existence, before that time did not exist. Our universe could not exist without time, and time could not exist without the universe, they are different components of the one entity.

According to Einstein's theory of relativity, time is regarded as a fourth dimension, on an equal footing with the familiar three dimensions of space. Einstein says that you can imagine all of space and time represented as a four dimensional space-time map, on which all of history, the present and the future of the universe can be represented. The four dimensions of space and time are collectively referred to as the space-time continuum, which by the way, is not just an invention from the script writers of Star Trek. The problem we have is in trying to visualise these four dimensions because we can only see the three dimensions of space, we cannot 'see' time. However, even though we cannot see it, it is necessary to include time if we are required to define a precise location. We can, for example, define an object's position in a room by three simple measurements, such as how far forward, how far to the left and height above floor level. These co-ordinates will define where the object is, but only where it is now, it may be somewhere else tomorrow.

Let's take a closer look look at the implications of time existing as a dimension. Imagine an object, not a quantum particle this time, or a massless photon moving at the speed of light, but an ordinary day to day object; let's use a one ton boulder. Now this boulder is behaving itself, as only boulders can, and isn't doing anything, its just sitting there existing. It's real, solid and big, there is no doubt at all in our minds that it is actually there, we can walk up to it and kick the thing. The question is, how long is it there for? Now at this point I should mention that this is neither the time nor place for a Smart Alec geologist to come along clutching his PhD and inform us that it is so many million years old and will remain in that condition for so many million more years. There is an even chance he would end up under it! We want to know how long in time it is there, in that spot, in front of us. We have a pretty good idea, we know it is less than a Femtosecond, which to remind you is one thousandth of one trillionth of a second. So after this very brief moment of existence in our time frame, what happens to it? Does it simply disappear when it slips from the present into the past? Equally where does it come from? Does it just spring into existence into our time frame? As time flows past us and the 'now' becomes history, has that moment gone forever, or does the past, and the future, actually exist independently of our own personal time frame?

Again, according to relativity, the past and the future, as well as the present, all exist equally. As I said earlier, 'you can imagine all of space and time represented as a four dimensional space-time map, on which all of history, the present and the future of the universe can be represented.' This would mean that our boulder isn't only existing in front of us in our personal time frame, it would be existing in *every* time frame, along with everything else of course. We can try to visualise this the following way. Imagine our boulder is resting on a glacier that over a period of 5 thousand years is slowly moving, and we want to show this

Keith Mayes

movement over time. We could make a jelly cube to represent time (I bet scientists have a lot of fun). The bottom would represent the past, at the beginning of the period we are interested in, and the top would be the finish, or the present. The length of our cube could represent a north/ south movement at an appropriate scale, and likewise the width could represent an east/west movement, and as stated, time is represented by height. We could slice through the cube at different heights to represent different slices of time.

To show the movement of the boulder over time we simply have to poke a rod through the jelly cube. If the boulder moved at an uneven rate we would have to go high tech and use a bendy wire instead.

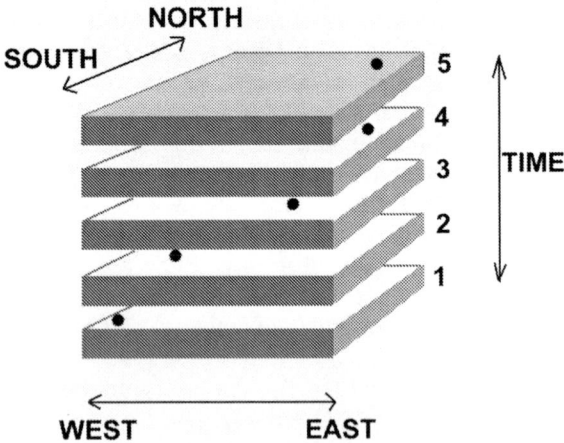

We can now slice up our cube into five slices to show us where the boulder was at 1000 year intervals. We could of course keep slicing our cube into ever small fractions showing ever more precise movements over ever smaller periods of time. Viewed this way, we can see that the existence of our boulder is continuous, it exists like a 'worm' through time, it only looks like a boulder at each 'slice' of time. We can make these slices as small as we like, but we will always only see the boulder as it is at that slice, or more accurately, at that moment in time. In this manner we can see that time does not 'flow' at all, it is a fixed dimension just as like the three dimensions of space, it is us that is moving through it.

The direction that we move through the dimension of time would appear to be governed by the arrow of time, as already described. As for the speed at which we travel through time, that is governed by the speed we are moving through the dimensions of space, as described by relativity.

If we put all this together, the picture we get is that there is only one speed we are able to travel at, and that is the speed of light, and it is a combination of our speed through space *and* through time. The faster we travel through the dimensions of space, the slower we travel through the dimension of time, and *vice versa*. Thus an astronaut zooming along at light speed has used up all their speed 'allocation' in the space dimensions, and as a consequence does not travel through time. This would seem to suggest that the speed of light really is the limiting speed within the universe, and if we had no motion at all through space then we would be travelling at light speed through time. In terms of our diagram, the more movement we make across the cube (space) the less we make in the direction of height (time).

If we use the same analogy of moving through our cube to describe moving through time, we can look again at the consequences of the globe-trotting clock. We will repeat the experiment using not only clocks this time, but twins as well, and we shall call them Bill and Ben. We send Bill on an exciting journey into space with an atomic clock, travelling at high speed. Ben stays at home, polishes his clock every day, and sorts his CD's into alphabetical order. When Bill returns home his clock of course will be running behind the time as shown on Ben's clock, and also will not be as shiny. We understand why this is so now, (especially the shiny bit) but what are the implications for the twins regarding their ages? Before Bill's journey they were the same age, but now Bill is a little younger than stay-at-home Ben. How can this be when they are still both sharing the same time frame? This is where we return to our jelly cube to make the situation easier to understand. Both twins will have their own time lines represented by their holes through the jelly cube. Due to Bill's travels he has moved across the space dimension a great deal more than Ben, and so has made less progress up the cube, which translates to less progress through time. In order for them to be still sharing the same time line, (slice) *and* have a difference in age, all we have to do is instead of cutting our slice of time parallel to the cube, as we have been doing, is cut it an an angle so that they both appear on the same exposed plane. To put this a little more scientifically, we could say that our movement through the space dimensions alters the *angle* at which we intersect the time dimension. I should clarify at this point that this is merely a representation to show the relationship between speed and time, and that the correct interpretation of relativity states that movement through space reduces our *movement* through time.

Another point that may need clarifying is defining movement through the space dimensions. In our example with Bill and Ben, although Ben stayed at home, he is of course still moving through the universe at high speed. The Earth is spinning on its axis while it orbits the Sun, which in turn is wheeling around the galaxy, which itself is flying through space, and so on. The movement we are referring to is the *relative* movement between objects that we are using to compare time.

The diagram of the jelly cube that we have used is only a schematic of course, as it is not possible to draw a representation of four dimensions. For the purpose of a visual representation I have had to 'steal' the dimension of height and use it to represent time. In reality the dimension of time is described as being at right angles to the dimensions of space, a concept that is impossible for us to imagine. Hopefully though, our jelly cube has served a useful purpose.

Although the model we have used describing time as another dimension agrees with relativity, it does have some unfortunate consequences, it says that the future is already determined! I dislike the idea that the future may be 'already there' just waiting for us to pass over it, because if true, where does that leave destiny and freewill? (See Do we have free will?) If on the other hand the future is not already there, then where does that leave the special theory of relativity? If time is 'real' in the same way that the other three dimensions are real, this would indicate that it does exists and is there waiting for our consciousness 'to move over it'. How much flexibility that allows for the future I do not know, but I must confess I don't like the sound of it.

I prefer to imagine time as being like our jelly cube, a fixed dimension that can be sliced up at any angle, but an 'empty' dimension, a blank canvas waiting for us to put our mark on it. I just do not like the idea that just because time exists as a dimension, all events *in* time also have to exist.

The truth of the matter is yet to be revealed, but personally, I have trouble in accepting the idea of the future being fixed, I believe that I exercise free will.

Acknowledgements:

John Gribbin *"Companion to the Cosmos"*, Paul Davis *"The Mind of God"*, J. Bronowski *"The Ascent of Man"*, Stephen Hawking *"A Brief History of Time" and "The Universe in a Nutshell"*, Timothy Ferris *"The World Treasury of Physics, Astronomy, and Mathematics"*.

06. Is time travel possible?

"The Moving Finger writes; and, having writ, Moves on: nor all thy Piety nor Wit Shall lure it back to cancel half a Line, Nor all thy Tears wash out a Word of it."
Edward Fitzgerald, 1809-83, in his poem 'The Rubßiyßt of Omar Khayyßm.'

Time travel is a concept much loved by science fiction writers, and Star Trek fans, but is it possible within the known laws of physics? To answer the question we shall take another look at quantum theory, bringing in the ideas of Richard Feynman, and re-examine the theory of relativity. After that we shall consider the difficulties involved in constructing a time machine, if such a thing can be done. But before we do any of this, let's first consider some of the problems that time travel raises. I have to say at this point that if ever a theory was riddled with problems, paradoxes, speculation and possibilities, then this is it, outright winner of the 'Science, the Universe and God' gold medal award!

One of the stock answers to the question of time travel is to suggest that if time travel were possible it would have already happened. By way of explanation let's say that at some time in the future a clever scientist invents a time machine that can travel through time in any direction, just like the fictional time machine created by H. G. Wells. So where is our intrepid time traveller? History shows that no time traveller has ever visited us from the future, therefore it is never going to happen. This appears at first glance to be a good solid argument against the possibility of time travel, but the argument is flawed. It may be that time travellers from the future have visited us, but have not revealed themselves in order to avoid changing the future. Alternatively, time travel may be possible, but only into the future, the past already having been determined. Finally, time travel into the past may be possible, but only into a different (alternative) universe, thus avoiding paradoxes.

The fact that we have no record of having been visited by time travellers does not exclude its possibility. Having dealt with that little problem we can now move on the fun part, paradoxes.

Paradoxes

This is where we start to run into some of the problems posed by time travel. The most commonly posed paradox is known as the 'grand parent' paradox. This states that if you could travel back in time you could murder your grand parents and thus prevent your existence, thus rendering it impossible for you to have gone back in time and killed them... no need to draw you a picture. Even more to the point perhaps, you could travel back in time and kill the person responsible for time travel before they discover it! However, popular though the grand parent paradox is, it only reveals the tip of the iceberg. Let's examine a theoretical time travel situation in more detail in order to highlight some of the problems involved.

Imagine that today you travel back in time to August 2001 and warn the authorities that the World Trade Centre in New York is going to be attacked on September 11th. They take the necessary action and as time unfolds the disaster is eventually averted. You are still in August 2001 at this point and the disaster has not yet been prevented, but it will be eventually because of the chain of events that you have put into motion, and you now wish to return to the time you came from. Here comes the Big Question - *can* you return to where you came from? Where you came from the attack *had* taken place, therefore where you came from no longer exists! If you do return to your original time, the World Trade Centre will still be standing (because you prevented the attack), so it *cannot* be where you came from. You will have changed the course of history, not just for yourself, but for the entire world. The question is, *will* the World Trade Centre still be standing when you return to your starting time? This is a major point of contention. According to one theory it both will *and* it won't! This is because your actions will have created an alternative possibility and in the process an 'alternative' universe, where we now have one universe with the World Trade Centre intact, and one with it destroyed, and we will all exist in both. The theory is firmly rooted in quantum theory that states that ALL alternative outcomes are possible. However, there is no evidence at this stage that the strange phenomenon found in quantum theory can be applied to the larger world.

Yet another theory suggests that alternative universes will not be created because events will conspire against you that prevent you (as in this scenario) from warning the authorities. This 'theory' it isn't really a theory as such, more an expression of a desire, because it is believed to be impossible (and very bad manners from the point of view of historians) to change history. The theory has no explanation as to how this 'prevention' could actually come about and has no solid theoretical basis.

The same arguments can be applied to travel into the future. Suppose I travel 24 hours into the future, armed with a loaded gun, (just in case, you never know) and happen to meet myself sitting at my computer, (can you meet and interact with yourself?) also armed with a gun, which would be most unusual for me as I do not own one. I am alarmed to see that the gun is pointed at me and warn my future self to put it down or I will shoot him. He prepares to fire the gun so I shoot him dead at the computer, then return. On my return, when tomorrow comes, I am sitting at my computer and receive a visit from 'me' from my past. He warns me that he will shoot me. I of course knew this would happen and took the precaution of keeping the gun beside me and aimed at the spot at the time I knew 'he' would arrive, and I shoot him first. Can I do that? That isn't what happened in my future, it was the 'me' at the computer that died. Even if I don't manage to shoot 'him' in time and still die at my computer, which one of us exactly is it that is still alive? We have created a nonsense scenario. Now this may sound like nonsense, and perhaps it is, but if the future really does exist then it does raise the prospect of meeting and interacting with yourself.

If we accept the idea that the future does exist, because time is just another dimension as discussed in the previous section, then these paradoxes would be unavoidable. If however, we take the view that only the dimension of time itself exists, and not the events within it, then this would suggest that time travel is impossible. Why? Because if future events do not exist, then it would be impossible to travel into the future to witness them, there would be no 'future' to visit.

We either have to accept that if time travel is possible we will by our actions create all sorts of paradoxes, or accept that we will create alternative universes. The paradox problem will not go away. As for alternate universes, the mind boggles!

Alternate Universes

The idea of having alternate Universes very conveniently solves the paradoxes raised by time travel, but is it a serious possibility? It sounds a bit far fetched, like something out of a science fiction paperback, but try not to dismiss the idea out of hand, it has some strong supporters.

The alternative universe theory (also known as the multiverse) suggests that for every possible outcome of an event, an alternative universe is created, with the result that somewhere out there I won the lottery last week, and so did you. If only we knew how to get there! This theory has the problem that there must be an infinite number of alternative universes existing to cover every possible outcome of every event. Which one, if any, is the 'real' one, or are they all equally 'real'? Furthermore, where does all the energy and mass come from that creates all these alternative universes? To try and find answers to these questions we have to return briefly to quantum theory.

We learned from the cat-in-the-box thought experiment that prior to observation the cat is in, what is termed in quantum speak, a 'superposition of states'. In plain talk, it is *either* dead or alive until the moment of observation, its fate is not determined beforehand. The explanation given for this strange state of affairs was the behaviour of the electron's probability wave spreading through first the box, then the room, and finally collapsing into an electron at the moment of observation. A different approach however, is to suggest that at the moment of observation *both* (or all, as the case may be) possibilities become realities. This is achieved by the creation of an alternate universe, in which the cat is in the opposite state to the cat in our universe. A refinement of this theory suggests that there are always two universes involved, but prior to the experiment they are identical in all respects. If three different experimental outcomes are possible, then we would have three initially identical universes, one of which would change. In general, we would need an infinite number of universes to cover all possibilities.

To quote from *'The Mind of God'*, an excellent book by Paul Davies: *'Now imagine extending this idea from a single electron to every quantum particle in the universe. Throughout the cosmos, the inherent uncertainties that confront each and every quantum particle are continually being resolved by differentiation of reality into ever more independently existing universes. This implies that everything that can happen, will happen. That is, every set of circumstances that is physically possible (though not everything that is logically possible) will be manifested somewhere among this infinite set of universes.'* I should add that Davies dislikes the 'many universes' theory.

We have looked at some of the problems associated with time travel, it is now time to look at the laws of physics.

Quantum theory
We shall now study another aspect of quantum theory, one that is directly related to time travel. I am going to borrow from the work of Richard Feynman, as he was the acknowledged master of clear explanation in physics. I shall be taking an example from his book *'QED The strange theory of light and matter'*. In this chapter of his book Feynman has been discussing how electrons and photons interact.

'Let's look at another event now. We begin with a photon and an electron, and we end with a photon and an electron. One way this event can happen is: a photon is absorbed by an electron, the electron continues on a bit, and a new photon comes out. This process is called the scattering of light. When we make the diagrams and calculations for scattering, we must include some peculiar possibilities. For example, the electron could emit a photon before absorbing one. Even more strange is the possibility that the electron emits a photon, then travels backwards in time to absorb a photon, and then proceeds forwards in time again. The path of such a 'backwards-moving' electron can be so long as to appear real in an actual physical experiment in the laboratory. The backwards-moving electron when viewed with time moving forwards appears the same as an ordinary electron, except it's attracted to normal electrons - we say it has a 'positive' charge. For this reason it's called a 'positron'. The positron is a sister particle to the electron, and is an example of an 'anti-particle'. This phenomenon is general. Every particle in Nature has an amplitude to move backwards in time, and therefore has an anti-particle. When a particle and its anti-particle collide, they annihilate each other and form other particles.'

Many thanks to Feynman for that explanation, I felt it best just to quote it rather than condense or try to simplify it, as I couldn't.

We have now learned that time travel is not only possible, it is a perfectly normal phenomenon, at least for particles. But is it possible in the macroscopic world?

Einstein's theory of relativity

Black holes, according to relativity theory, warp spacetime with their enormously powerful gravitation field. The effect of this gravitational field is that if an astronaut were to cross the event horizon of a black hole, time would slow down on board his spacecraft as he approached the singularity and eventually come to a stop. Similarly time slows down in proportion to speed, the faster our astronaut travels the slower time runs. The closer the astronaut travels to the speed of light the more time slows, until at the speed of light, time would stop. Both these effects of time being affected by speed and gravity have been discussed in the previous section, all of which illustrates that time is not a fixed constant, it is affected by gravitational fields and relative speed in the same manner as the other three dimensions of space.

The solutions to particular equations of the Special Theory of Relativity can be expressed mathematically in any direction of time without running into any problems. Does this mean that time travel is possible? There is nothing in relativity that rules out time travel, it would appear to be *theoretically* possible.

Constructing a time machine

Research carried out in the late 1980's showed that genuine time travel is not forbidden by the known laws of physics. This means that it may be possible to build a time machine, but not that it may be easy. Help may be at hand though, it is possible there are naturally occurring objects in the universe that act as time machines.

There are at least two ways to build a time machine. Frank Tipler published a possibility in the highly respected journal *Physical Review* in 1974. This involves making a naked singularity. This is a singularity that is not concealed from view behind the event horizon of a black hole. To make a naked singularity involves rotating a singularity extremely rapidly, and if rotated sufficiently fast it would fling away the event horizon and exposes the singularity. We know that spacetime is extremely distorted by the singularity's strong gravitational field and the effect of this rotation would be to twist spacetime, and tip it over so that one of the dimensions of the space dimensions is replaced by the time dimension. A carefully piloted spaceship taken close to the singularity would enter the time dimension and journey through time instead of space, although to the astronauts all would appear as normal. When the spaceship moved away from the distorted area around the singularity, it would be in a different time from when they had entered the area.

According to Tipler's calculations, the same effect could be achieved with a cylinder about 100 km long and about 10 km across, made of material compressed to just over the density of a neutron star, and rotating twice every millisecond. It would be like ten neutron stars joined pole to pole and given a strong twist. Curiously, there are objects in the universe which nearly fulfil the other requirements - so-called millisecond pulsars are known which contain almost the right density of matter and spin once every 1.5 milliseconds, at one-third the speed needed to make a time machine. Such objects are so close to being time machines that they hold out the tantalizing possibility that an advanced civilisation might be able to tweak them up in the right way to allow time travel.

That such things as naturally occurring time machines exist in the universe, with only a little tweaking needed, raises the prospect that an advanced civilisation may have already done the trick! This raises the interesting possibility that such a civilisation would have the capacity to travel between the galaxies; a journey of a few million light years would be as nothing. Something for the UFO brigade to mull over!

The other possibility for building a naturally occurring time machine involves worm holes - tunnels through spacetime which may, according to relativity, connect a black hole in one part of the universe to a black hole in another part of the universe. Before the mid-1980's physicists believed that such objects as wormholes could not 'really' exist, and that a better understanding of Einstein's equations would prove this. They were forced to change their minds as a result of careful investigation of wormholes carried out by Kip Thorne and his colleagues at Caltech in 1985. It is interesting to note that this research was triggered by Carl Sagan, a well known scientist, who was writing the science fiction novel *'Contact'*, a best seller that went on to become a highly successful film. Sagan wanted his wormhole to be as scientifically accurate as possible and approached Thorne to check out the idea as presented in the book. What neither Sagan nor Thorne first realised from the results of Thorne's study was that this short-cut through space would also work as a short-cut through time. In 1988 Morris, Thorne and Yurtsever (Morris and Yurtsever were students of Thorne) published their conclusions in the journal *Physical Review Letters*, that Einstein's equations really did allow for the existence of wormholes that link different times, and could be used as time machines. (For further reading on constructing a time machine and time travel I recommend John Gribbin's book 'Companion to the Cosmos' from which much of this particular section was sourced).

We have seen that the laws of physics do not preclude the possibility of time travel, and further, that it may be possible to construct a time machine by tweaking naturally occurring objects in the universe. It would appear that we only need the technology to make time travel a reality.

However, regardless of the theories, I cannot believe that time travel will ever be possible. My main reason for this is because we have recorded history and we know what events happened in the past, and in our own recent past we even have our own memories of past experiences. If time travel were possible then we could not have a recorded history because it would be constantly changing. I can remember, for example, the day when Neil Armstrong set foot on the Moon, the first man to do so. If time travel were possible it would be possible to travel back through time and prevent that event from taking place. It would, for example, be possible for a Russian to travel back through time armed with the technical knowledge necessary to land on the Moon and give it to the Russians years before the Americans achieved it. That act would change history. It would then be possible for an American to travel back in time and prevent the Russians from using that information, and so on, and so on. History, however, is not subject to change, it remains constant, if it did change then it would

not be history! In order to have a past that past must be unalterable, if it were subject to change then we may not be born to observe it, but we are here observing it. The following quote sums it up rather well I think - *"This only is denied to God: the power to undo the past."*
Agathon (448 BC - 400 BC), from Aristotle, Nicomachean Ethics.

It has been argued that perhaps time travellers from the future have visited us, but being aware of the dangers of interfering with the past - that could have consequences for their future - they only observe in secret and do not interfere. However, it should be clearly understood that this policy of 'non-interference' does not help the situation at all. Imagine that today no time travellers have yet come back from the future to visit us. Now imagine that at some future time they do come back to today and visit us. Even though they may only be observing for a few minutes and then return to their own time, *they have still changed the course of history.* In the first instance no time travellers had visited us, but in the second instance they had. This alone has changed history, it doesn't matter whether or not anybody knew they were there, or if they physically changed anything or not, the fact is they were there. So what happened to our 'original' history, the 'today' with no visiting time travellers? It has been removed from history and never happened - but the problem is we know it did happen! Does this mean that when history is changed our memories are changed as well? Perhaps I only think that my memory of Armstrong walking on the moon has never changed, but in reality it may have been changed many times. I don't think so though, because it is recorded history, it is written down in black and white, and how could that change? Unless it has changed every time of course and I only think it is permanent? This is leading us on to the alternative universe theory, as it is beginning to sound a lot like it.

The alternative universes theory was designed to overcomes the problems of altering history, which of course cannot be altered. It is my opinion however, that this theory is an extremely complicated concept only offered up as a way of getting round a very real problem - that of changing history and associated paradoxes - and is based on strange phenomenon found only in the quantum world - which is not at all understood - and for which we have absolutely no evidence for being applicable in the larger world. Furthermore, I have yet to find any explanation of where all the matter and energy would come from that mysteriously creates all these very convenient alternative universes. I am also puzzled as to what would constitute a choice of outcomes that would generate a universe for every possible outcome. I can understand a photon being 'forced' by observation to go through either one slit or another, and this generating two possible outcomes both of which require its own universe, but what of other examples of a choice of outcome? What about my turning left at a road junction instead of right, does that create an alternative universe? Or what if I said 'yes' instead of 'no' when asked if I liked the colour pink, would that create an alternative universe? You can see where I am going here, if every choice resulted in an alternative universe, then the number of alternative universes created would be infinite and ever increasing at an infinite rate as all those alternative universes created their own alternative universes, and so on *ad infinitum*. All this just to avoid paradoxes! I think not.

I really cannot see how time travel could ever become a reality, for if it were possible it would have already happened, if you see what I mean?

Acknowledgements:
Paul Davies *"The Mind of God"*, John Gribbin *"Companion to the Cosmos"*, Richard Feynman *"Six Not-So-Easy-Pieces"*, Robert Gilmore "Alice *in Quantumland*". Richard Feynman *"QED The strange theory of light and matter"*.

07. Can we travel faster than light speed?

"Not only is the universe stranger than we imagine, it is stranger than we can imagine."
Sir Arthur Eddington (1882 - 1944)

In order to journey to the stars we will need to travel much faster than is currently possible with existing technology. Even at the amazing speed of light, at 186,000 miles per second, (670 million mph) the journey to our nearest star, Alpha Centauri, would take 4 years. To be accurate, it would take even longer than that because of the time it would take to reach light speed - accelerating at a rate that would produce 2G's force it would take six months, and of course another six months to slow down. To reach our nearest neighbourhood spiral galaxy, the Andromeda Spiral Galaxy, would take over 2 million years. We need to go even faster than light, but is it possible?.

Relativity tells us in no uncertain terms that faster than light speed is impossible, period. This is not because of any technical restrictions (as if that wasn't enough!) but simply because it is a fundamental part of the way spacetime is constructed. Let's look at the reasons why.

Mass and Energy

Imagine I challenge you to a rock throwing contest to see who could throw a rock the furthest. There are a selection of rocks to chose from, some big heavy rocks and a few smaller lighter rocks. It is obvious we would both choose one of the smaller rocks to throw, because they are lighter. We both pick our small rocks and make our throw, and much to my annoyance, you win.

When we are throwing our rocks we are trying to accelerate its speed as much as possible before letting go, knowing that the faster it is travelling the further it will go. The reason that you won is because you managed to accelerate the rock more than I did. To put this on a more scientific footing, we can say that the lighter rock has less mass, and that with a given amount of energy an object will travel faster - and therefore further - the less mass it has.

However, I was unhappy with the result of the rock throwing contest as you are much bigger and stronger than I am, which made it an unfair contest. In effect, you were using a bigger 'engine' then me to accelerate the rock. In order to overcome the problem of our different 'engines', I challenge you to a car race. The car that achieves the fastest time around the race circuit wins the contest. We both agree that this is a much fairer contest as we are now evenly matched - providing our cars are identical.

We are both presented with cars having the same size engine so the winner of the contest should be the driver with the better driving skills. We are both given identical big gleaming powerful cars to drive at high speed (I wish). What you don't know is that I was so upset at loosing the rock throwing contest (being a poor looser) I have cheated this time in order to ensure my victory and have bribed one of the mechanics to secretly place 1000 pounds of lead into the boot of your car. (cackle. cackle). The race duly takes place, and surprise, surprise, I am victorious! My celebrations are short lived however as an official enquiry soon reveals my dastardly scheme. How did you know that I had cheated? It is because although both cars had the same size engine and should have been identical in every respect, your car ran out of petrol, while mine didn't. Furthermore, when driving flat-out down the straight, my car went faster then yours, which it should not have been able to do. Why? Simple isn't it, your car was a lot heavier than mine, and given the same size engine was not only unable to travel as fast, but used more petrol in the attempt. As in the rock throwing contest, we can explain this in more scientific terms. We can say that the more mass an object has the more energy it requires to reach a given speed, and to maintain that speed.

We now know that in order to achieve the highest possible speed we need an object with the lowest possible mass, with the most powerful possible engine to accelerate it. There is something else we need to consider though, and it concerns relativity.

The Special Theory of Relativity

Relativity tells us that as an object increases in speed, so it increases in mass. This sounds impossible, how can an object increase in mass? The answer goes to the very heart of relativity, to Einstein's most famous equation: $E = MC^2$. In plain talk this tells us that energy and mass are interchangeable, and we have witnessed this effect in the detonation of atomic bombs. We do not though, need to understand the workings of an atomic bomb in order to understand the principles involved, but it is a little complicated. The special theory of relativity tells us that the mass of an object increases with high speed. This mass increase is not detectable to anyone travelling with the object, only to observers in the frame of reference where they see the object to be moving. For them, the measured mass of the moving object increases in accordance with the equations of relativistic mechanics until, if the object could reach light speed it would have infinite mass.

In case you are wondering - the 'E' is for energy, 'M' for Mass and 'C' for light speed. In other words, energy equals mass times the speed of light squared.

The Lorentz contraction.

There is another effect of relativity for objects travelling at relativistic speeds. Apart from gaining mass, they also shorten in length in the direction in which they are moving - they get smaller! Let's look at another everyday example to make sense of this.

We shall imagine that we have just bought a brand new sports car. So excited are we with our new purchase that we immediately drive to the nearest race track to put it through its paces. On arrival we cautiously decide to let the track expert test the car first. He sets off round the race track and very soon is travelling at 120 mph. We are delighted with the car's performance but start to worry that it may be too large to fit into our garage, and decide to calculate the cars length as it passes us by. We know the speed is 120 mph, so we set up a very accurate device that times precisely how long it takes the length of the car to pass a fixed point. This imaginary measuring device is so accurate it can measure to within an accuracy of 13 decimal places. It is then a simple matter to multiply the speed of 120 mph by the time taken by the car to pass our measuring device to arrive at its length. The answer comes out at 15.9999999999974 feet. When the car eventually comes to a stop we decide to check our measurement and find that the car is precisely 16 feet long. There is no fault either with our measurement of the car at speed or at rest, the difference is real. In practice we would not be able to detect this tiny difference at such a low speed relative to the speed of light, but at far greater speeds the difference becomes very noticeable. If a spaceship were to travel at 580 million mph (about 87% of the speed of light) the length of the spaceship would be half that of when it was at rest. Why is this?

The effect is known as the 'Lorentz contraction'. You will recall that in the section 'What is time?' we examined the effect that speed has on time, and how the faster we move in space the slower we move in time. This effect is known as 'time dilation'. If we now apply this knowledge to the measurement of our speeding car, we can see why it gets shorter as it goes faster. It is necessary to take into account at this point that all movement is relative. We cannot say for example that a star is moving at 1,000,000 mph, we have to say what it is moving at this speed in relation to. It may be moving away from us at that speed, but may be moving at 5,000,000 mph in relation to another galaxy. In our example with the moving car, the speed of the car is in relation to us and our measuring device. Relativity tells us that it makes no difference which of us is actually moving (we both are when you take into account the rotation of the Earth).

Let's take the example of a high speed space ship whizzing past the Earth. From the perspective of the astronaut onboard he is stationary while the Earth rushes by, and hence sees our clock as running slow. As a result he realises that our indirect measurement of his spaceship will yield a shorter result than when it was measured at rest, since in our calculation (length equals speed multiplied by elapsed time) we measure the elapsed time on a clock that is running slow. If it runs slow, the elapsed time we find will be less and the result of our calculation will be a shorter length. This is an example of the general phenomenon that observers perceive a moving object as being shortened along the direction of its motion.

We can now pull all this information together into one neat package, containing The Special Theory of Relativity, the Lorentz Contraction and Time Dilation. Sounds impressive doesn't it! We can describe our neat little package with the following statements:

1) Speed results in an object gaining mass
2) Speed results in an object shortening along the direction of its motion
3) Speed results in time slowing down

It is worth noting at this point, that the fastest moving particle in the universe is the photon, the particle of light, as it has zero mass. Although Einstein's equations tells us that nothing can be accelerated to, or beyond, the speed of light, the photon *exists* at light speed, it does not accelerate to it. As a further point of interest, Einstein's equations do not rule out the possibility of particles that exist at faster than light speed, which would result in them being unable to travel at *less* than light speed.

Armed with all this useful information let's set of on an exciting high speed journey through space in an attempt to travel at faster than light speed (as we have finished sorting our CD's into alphabetical order). We set off, and as we journey through space we gradually increase our speed until we have reached 98% of light speed. At this point a stationary observer would view our spaceship as being 80%

shorter than if it were at rest. Our density has increased and time is running slow, not that any of these effects would be apparent to us. We continue to accelerate (this is the Grand Prix GTi version complete with go-faster stripes and a catchy bumper sticker) and reach 99.5% of light speed. At this speed time aboard our spaceship is running at only one tenth that of a stationary observer. We continue to increase speed until we are at 99.9% of light speed. Time is running extremely slowly, our length reduced drastically and our mass has increased 22 times. Our increase in mass results in us having to boost our engines to compensate. We still push on towards light speed and reach 99.999% of light speed. Things are really hotting up now. Our mass has increased by a factor of 224, time has almost stopped, and our size reduced to a mere dot! We continue to accelerate and reach 99.99999999% of light speed - almost there, but things are getting really difficult now. Our mass is now increased by a factor of 70,000, our clock is moving so slowly that it appears to have stopped, and our size is reduced to almost nothing!

At this point we have to give up trying to go any faster. As our speed continues to increase so does our mass, and therefore the amount of energy required to accelerate. Our mass will continue to increase, and as we approach light speed it would begin to approach infinity and the energy required would also approach infinity. At light speed, if it were possible, our mass would be infinite, we would require infinite energy, our size would be infinitely small, and time would stop.

We cannot travel at light speed. The speed of light appears to be an integral part of the nature of space-time, and as discussed in 'What is time?' it would appear that there is only one speed that we can travel at, and that is the speed of light as a combination of our speed through space *and* through time.

Now here is the 'problem' part that you knew was coming, and it comes once again from the strange world of

Quantum Mechanics. We just don't seem to be able to escape from it, do we?

The EPR experiment.

This experiment was devised by Albert Einstein, Boris Podolsky and Nathan Rosen (hence the 'EPR') as a thought experiment to 'prove' that quantum theory was incorrect. The technology did not exist then to actually carry out the experiment but Einstein believed, that in principle, it proved the 'foolishness' of quantum theory. The experiment was designed such that it would result in communication at faster than light speed, which Einstein's theory of relativity showed to be impossible. Neils Bohr's theory of quantum mechanics was at odds with Einstein's relativity, because it allowed instant communication between paired particles.

We have seen from our previous examination of this experiment (What is Quantum theory?) that it has actually been carried out over a distance of 10 kilometres and confirmed as correct, instant communication did take place. Once again, we have a situation where, in the quantum world, the 'impossible', can take place, this time in the form of instant remote communication, which does of course mean that communication is taking place at faster than light speed.

Can light travelling at faster than light speed?

In a paper dated 19 July 2000 A team of scientists announced that they had succeeded in sending a pulse of light through a special chamber at a velocity faster than the speed of light. Scientists from the NEC Research Institute in Princeton, New Jersey, explain how they sent a pulse of light through a six centimetre chamber containing an unnatural form of cesium at the even more unnatural temperature of nearly absolute zero. The pulse of light travelled so fast that its peak actually exited the cesium chamber slightly before it entered. *"No intuitive way to explain this observed effect precisely can be found because the 'specially prepared' atomic cell (cesium chamber) is in a state that does not exist naturally,"* write researchers Lijun

Wang, Alexander Kuzmich and Arthur Dogariu in a statement. The team is quick to point out that their work does not violate Einstein's Theory of Special Relativity, which states that nothing can travel faster than the speed of light, because this would entail going backward in time. *"More or less you can't go faster than the speed of light,"* said Wang. But the restriction that applies to things made of matter does not apply to light waves.

In fact, it was by using the waves of different colours of light to amplify each other and create the pulse that the researchers were able to get the light to warp through the cesium cell and reconstruct itself on the other side before it had entered. According to the researchers, the experiment also does not violate the principle of causality, which requires the cause of any effect to precede it in time. The fact that the peak of the pulse of light exits the chamber before it enters is the result of the light waves building a pulse on the other side of the cesium cell that is identical to the one entering it. So it is not exactly the same pulse. *"This means that even if the 'effect' appears to precede the 'cause,' you still can't send information - such as news of an impending accident - faster than (the speed of light),"* writes Jon Marango of the Imperial College in London, in a commentary on the work also appearing in Nature.'

It would appear that, perhaps, faster than light speed is possible, sort of, for light. It does not however allow for any form of communication, apart from one particle telling its other paired half what state to collapse into, and certainly does not allow it for solid matter; or so it seems so far, but things have a habit of changing.

Acknowledgements:
John Gribbin *"Companion to the Cosmos"*, Davidson & Smoot *"Wrinkles in time"*, Michio Kaku *"Hyperspace"*, Stephen Hawking and Roger Penrose *"The nature of space and time"*.

08. Is there extraterrestrial life?

"Sometimes I think the surest sign that intelligent life exists elsewhere in the universe is that none of it has tried to contact us."
Bill Watterson

Whether or not extraterrestrial life exists is of course open to speculation, for until we actually find it the answer has to be that we do not know, for not finding it does not preclude the possibility of its existence.

When considering the vastness of the universe, containing in all probability many millions of planets, it is difficult to imagine that our planet is the only one that harbours life. However, If that unlikely possibility should turn out to be the case, then it would perhaps be only reasonable to assume that it must be *impossible* for life to form by the process of natural development, but rather that it must have been uniquely created. To take the opposite view, that life formed spontaneously on only one planet in the entire universe, is pushing the laws of probability to virtually unlimited bounds. I would suggest therefore we have only two realistic options to consider. 1) Life was created exclusively on this planet by God. 2) Life has formed on many planets spontaneously.

Taking the first option, that life was created exclusively on this planet by God, raises a number of problems. One problem is the story of how God created human life as recorded in the bible, by forming Adam from the 'dust of the ground' (Genesis 2.7). Suffice it to say that this version of the formation of the species 'Homo Sapiens' does not accord with the findings of anthropologists, and I do not feel it necessary to embark on a lengthy discourse explaining why - we all know why. But in fairness, it has to be said that the bible was written in the style of the people the best part of 2000 years ago and should not perhaps be taken too literally today. We now live in age where man has walked on the Moon, has sent probes to all the planets in the solar

Keith Mayes

system, and has even sent probes to investigate comets and land on an asteroid. Armed with all this knowledge that we have gathered about other worlds and distant galaxies, the vast majority of us no longer believe that the world was made in six days, so I will not pursue the pointless exercise of arguing against the bible. Those of you that do believe in the literal truth of the bible have your own reasons for doing so, but I do not believe that God created life on Earth. I do not argue that God did not create the universe which *resulted* in life on Earth, but that is not the same thing at all.

If then we discard the idea of God creating life on Earth we are left with the second option that life was created here by chance and therefore it is most likely that it will also form elsewhere. You may ask at how I reach this conclusion, and my reply is that it is based on probabilities. Let's look at how many planets there may be in the universe, starting with the number of known galaxies.

1) The number of galaxies. An estimated 50 billion galaxies are visible with modern telescopes and the total number in the universe must surely exceed this number by a huge factor, but we will be conservative and simply double it. That's 100,000,000,000 galaxies in the universe.
2) The number of stars in an average galaxy. As many as hundreds of billions in each galaxy. Lets call it just 100 billion. That's 100,000,000,000 stars per galaxy.
3) The number of stars in the universe. So the total number of stars in the universe is roughly 100 billion x 100 billion. That's 10,000,000,000,000,000,000,000 stars, 10 thousand, billion, billion. Properly known as 10 sextillion. And that's a very conservative estimate.
4) The number of stars that have planetary systems. The original extra-solar system planet hunting technology dictated that a star needed to be to close to us for a planet to be detected, usually by the stars 'wobble'. Better technology that allows us to measure the dimming of a stars brightness when a planet crosses its disk has now revolutionised planet hunting and new

planets are being discovered at an ever increasing rate. So far (August 2003) around 100 have been discovered so we have very little data to work on for this estimate. Even so, most cosmologists believe that planetary formation around a star is quite common place. For the sake of argument let us say it's not and rate it at only one in a million and only one planet in each system, as we want a conservative estimate, not an exaggerated one. That calculation results in: 10,000,000,000,000,000 planets in the universe. Ten million, billion, as a conservative estimate.

5) The number planets capable of supporting life. Let's assume that this is very rare among planets and rate it at only one in a million. Simple division results in: 10,000,000,000 planets in the universe capable of producing life. Ten billion!

For a more scientific approach I recommend The Drake Equation. This states that the number of *communicating* civilisations in our galaxy (note, our Galaxy only, not the universe) likely depends on a number of factors which must combine to yield a habitable planet where life has the chance to develop to a certain level of technological know-how. These factors include the rate of formation of stars like the Sun, the fraction of those with planets, the fraction of Earth- like planets, the fraction of such planets where life develops, the fraction of those where life becomes intelligent, the fraction of intelligent species who can communicate in a way we could detect, and the lifetime of the communicating civilisations. As you may imagine, There is a lot of debate about reasonable values for most of these factors.

Frank Drake's own estimate puts the number of communicating civilisations in just our Galaxy alone at 10,000.

Even though the figures I have used cannot of course be considered to be accurate, at least the figure of 10^{21} stars in the universe is most definitely an underestimate. The number of life supporting planets that may be orbiting those stars is impossible to say, but by any reasonable estimate must surely run into the millions, if not billions. This is easy to justify on the basis that following the Big Bang the most abundant material in the universe was hydrogen and helium, being the most simple atoms, and this material forms the bulk of the raw ingredients for star formation. All stars begin life in the same manner, by the gravitational drawing together of these basic elements that then gravitationally collapse to form a star. Apart from size, all stars begin pretty much the same, with the remnants of the hydrogen and helium clouds that are not absorbed into the stars forming an orbiting disc that goes on to form the protoplanets. With this same process repeated many billions of times it would be only statistically reasonable to expect that many planets would have similar characteristics, and would be capable of supporting life of one form or another, just as our planet does.

In order to answer the question of the existence of extraterrestrial life, it need exist on only ONE other planet. Given those odds, how can it *not* exist?

In view of the incredibly high probability of extraterrestrial life existing the question remains why we haven't yet detected any signs of it. Maybe there has not yet been enough time, or maybe we are using the wrong technology. I appreciate that there are a number of people that claim that aliens are already visiting us in the form of UFO's, but as that is very speculative I have dealt with the subject separately. See "UFO's and alien abductions'.

The SETI team (Search for Extra Terrestrial Intelligence) has been scanning the heavens for many years searching for a radio signal from an alien source, but so far without success. Having said that, they do have many very good candidate signals that they will follow up on. Radio signals do seem to be the logical way to send messages over astronomical distances, but an alien society with a mere hundred years advantage in technology could be using something very different. With a thousand years advantage, and a thousand years in evolutionary terms is the blink of an eye, they may have technology beyond our comprehension. It could be that their radio signals - if they ever used radio at all - were transmitted only briefly and may have reached us years before we had radio receivers, so we will never find a radio signal. Alternately, they may have started using radio more or less the same time as we did, so if they are thousands of light years away we will not receive their message for thousands of years to come, by which time we may no longer be using it. The window of opportunity for two distant civilisations to have compatible communication systems may be very small indeed.

We need to take into consideration the length of time it would take for a signal to reach us. Our own 'Milky Way' galaxy measures roughly 100,000 light years across, so a message from the opposite side of the galaxy would take 100,000 years to reach us. Our nearest spiral galaxy, the Andromeda Spiral Galaxy, is over 2,000,000 light years away. For a message to reach us today it would need have been transmitted towards the end of the Pliocene age, when Homo Erectus first appeared and began to diversify. To put it another way, the race that transmitted such a message would be 2,000,000 years more advanced then we are and we can only imagine how primitive we would appear to them. We need to consider also that if we did receive a message from the Andromeda Spiral galaxy, by the time we did they may no longer exist.

Even if there are many civilisations in our own Milky Way galaxy, and even if we do hear from them, do we really want them to know where we are? I for one would rather hide in the long grass, my mother taught me never to speak to strangers!

Update: August 2003

Astronomers have published a new estimate of the total number of stars in the universe. The international group of astronomers presented their findings at the General Assembly of the International Astronomical Union in Sydney, Australia. The figure they have arrived at is 70 sextillion, seven times higher than my estimate (hey, at least I got the right number of zeros!). This figure does not represent the actual total number of stars in the universe, just those that are in range of our telescopes. The actual number could be very much larger.

Acknowledgements:
I always try to list my main sources of information so that those interested can look up the books mentioned for further reading. The sources used for this section are mainly from stuff floating around in my head and from various magazine articles and the internet.

09. Asteroid Impacts. Can we survive?

"I do not know what I may appear to the world; but to myself seem to have been only like a little boy playing along the sea-shore, and diverting myself in now and then finding a smoother pebble or a prettier shell than ordinary, while the great ocean of truth lay all undiscovered before me."
Isaac Newton

Most of us have, from time to time, witnessed the occasional 'shooting star' streaking briefly across the night sky. These so called 'shooting stars' are only tiny grains of comic dust and debris burning up with friction as they speed through our atmosphere, and are more correctly termed meteors. A meteor the size of a garden pea, being larger than average, creates a beautiful sight as it blazes across the sky, and is bright enough to be easily photographed.

Periodically throughout the year, during its annual orbit around the Sun, the Earth will pass through a region of space that is densely populated with dust and particles. This is because our orbit has intersected with the orbit of a comet and passed through the old 'tail' of the comet, which is composed of billions of these tiny particles. When this happens astronomers, such as myself, will eagerly sit out in the garden from midnight onwards, in usually freezing cold conditions, to marvel at the spectacle of dozens, if not hundreds, of meteors putting on a celestial fireworks display. These events are well known in advance and eagerly anticipated each year, such as the Leonids meteor shower that reaches maximum on the night of November 17th as it passes through the trail left behind by the comet Temple-Tuttle.

Every once in a while we run into larger objects, the size of rocks, and these are large enough to survive passage through the atmosphere without burning up completely, and they reach the Earth's surface. These objects are called meteorites and can provide scientists with important information about the early formation of the solar system. If a meteorite is sufficiently large it will impact with enough force to form an impact crater, and many of these can be seen around the world.

Larger rocks that are orbiting the Sun are called asteroids, and these can measure many miles across, the largest known of which is Ceres with a diameter of 933 km (571 miles). It is estimated there there are over a million asteroids over 1 km across. When an asteroid impacts with the Earth it creates much more than just an interesting 'shooting star'.

Asteroid Impact

So what happens when the Earth is hit by a large chunk of rock? A 1 km asteroid for example, striking the Earth at a typical speed of 25 to 30 kilometres per second would have a devastating effect. The asteroid will accumulate enormous kinetic energy as it speeds towards the Earth, accelerated by the Earth's gravity. On impact this kinetic energy will be instantaneously released into the impact area rock in an explosion equivalent to 300,000 megatons of TNT - the largest man-made nuclear weapon had a yield of 60 megatons. The Flash and blast from the impact will destroy an area the size of the USA state of Arkansas. Within a few seconds A 20 kilometre wide crater will be excavated, and the resultant debris will be ejected into sub orbital trajectories. This debris will later re-enter the atmosphere like a massive meteor shower all over the planet creating an intense global heat pulse, raising fires that will destroy a significant proportion of the animal and plant life of the planet. The ozone layer will be completely destroyed. The shock wave from the impact will spread through the planet creating major volcanic and seismic activity. When the asteroid entered the atmosphere the ionisation of the air so caused would result in intense acid rain falling word-wide, together with large quantities of pyrotoxins produced by global fires.

The situation would not be any better if the impact occurred at sea. In addition to these effects, an impact at sea will produce a tsunami, capable of travelling considerable distances, and possessing enormous energy. Such surges will pose a substantial threat to low lying coastal areas. An impact in the Atlantic Ocean by a 1 kilometre asteroid will create a deep water wave 10 to 15 metres high. When it hits the continental shelf of Europe and North America, travelling at 600 kilometres per hour, it will run up a wave height of between 300 and 800 metres, depending on coastal topography.

The result of this initial impact will destroy a large percentage of life on Earth, but this is only the beginning. Those that survive this far may come to wish they hadn't. The vast amount of dust and debris injected into the upper atmosphere, combined with smoke from the firestorms will produce the greatest threat. The surface of the Earth will be shrouded in darkness and it is this that will pose the greatest threat to the global ecosystem as photosynthesis stops, food chains collapse and cold and starvation set in. Nothing will grow, animals will starve, we will starve. After a year, or perhaps two, the atmosphere will clear, but the Earth's albedo will be higher due to snow and ice, and it will reflect more of the Sun's radiation, leading to a runaway feedback situation, possibly leading to a new ice age.

All that from just a 1 kilometre wide asteroid! Statistically, we are hit by an asteroid of this size, not every 30 million years, but every 100,000 years.

We have had some near misses in the last few years. For example, at midnight GMT on August 10th 1998, asteroid ML14 crossed the orbit of the Earth at the exact point the Earth had occupied only 18 hours earlier. Had ML14 reached that point at 06.00 the previous morning, an area the size of France would have been totally devastated by 06.05. By 08.00 most of the world's vegetation would have been in flames. By late October 30-40% of the human race would be dead or dying. ML14 has a diameter of 2 kilometres.

In 1908 an impactor detonated some 5 kilometres above the ground in the Tunguska region of Siberia which yielded an estimated energy equivalent to 20 megatons of

TNT. A Tunguska sized impact over London would destroy everything within the M25. The impactor would have been in the size range of 50-100 metres and the statistical time-scale for such impacts is between 50 and 100 years.

It is not a question of if we will be hit by a mass extinction sized asteroid or comet, but when. It is going to happen. Whether or not the human race will survive the impact remains to be seen.

Defence Systems

What can we do to protect ourselves? One chance we have is that if we are given another twenty years or so, we could develop a missile defence system and use a nuclear blast to nudge the asteroid off course, providing we spot it in time. Talks are in place for such a defense system, but you know how it is with budget cuts, and how long talks can go on for.

It is not a simple matter to nudge an asteroid off course, it depends how much notice we are given, which depends on when we first happen to spot it and are then able to calculate its projected course. We may have anything from six years to six days notice. After it has been spotted we need to find out its size, which can be fairly accurately arrived at, and also its composition, which can't. We need to know what the asteroid is made of and how densely this material is compacted together. This information is vital if we are to successfully divert it from a collision course and be able to calculate the size of the explosion necessary, together with the detonation distance, to divert it but not break it up into many smaller pieces. If it did break up then the many pieces so created would still impact with us and the end result for our planet would be pretty much the same as if it hit us in one piece. In order to gain this information we would ideally need to land a probe on its surface, analyse the data transmitted back and hope that not only have we got it right but that we have the time and technology to do the job. It's asking a lot.

Another method that could be used is to land a powerful engine onto the asteroid and clamp it securely to the surface, providing the surface is stable enough to allow this. Firing the engine over a long period of time may be enough to deflect it just enough so that it avoids impacting with us. Again this would not be an easy task by any means.

A third method would be to simply strike it with an atomic missile that was so powerful it would vaporise the entire asteroid. Sounds easy, but again we would need sufficient time, a powerful enough warhead to guarantee total destruction, and the technology to accurately deliver it to a very tiny distant point in far space. Don't make the mistake of thinking that the technology already exists, because quite simply it doesn't. The United States still does not have its 'Star Wars' missile defence system yet, and that is child's play compared to the technology required to detect and destroy an asteroid. It must also be considered that the nuclear warhead would need be so massive to vaporise the asteroid that the risk of launching such a devastating bomb would be totally unacceptable as a launch failure would be catastrophic.

Personally, I do not imagine that we will have an asteroid impact defence system in place for at least another 50 years or more. Governments would rather use their money and technology on weapons of war than on a threat that most likely will not materialise during their term of office.

Let's all hope that we do not suffer the terrible destruction that a large asteroid would inflict on this planet while governments merely engage in talks on asteroid defence systems whilst actively engaged in developing weapons of mass destruction under the guise of 'missile defence systems'.

I can imagine there are lots of people thinking it could never happen. Why not? Did you not see comet Schoemaker-Levy smash into Jupiter in 1994? Had it hit Earth instead, we would not be here, not any of us. Some people think that such a terrible event simply could not happen, simply because it is so terrible. Perhaps they believe that someone is watching over us. Take a look around, do you think someone is watching over us?

I do not think we are here to serve a purpose, or that anyone is looking out for us, (God for example), we just happen to be here, oddly enough because of the asteroid that destroyed the dinosaurs. Next time it could just as easily be us. There are no guarantees. We're nothing special in the great scheme of things.

Can we asses the risk of Impact?

Figures are published by various sources from time to time indicating the probability risk of a major impact causing massive loss of life, referred to as 'Extinction Level Events'. These figures, guesses really, vary up and down with the weather and cannot be relied upon. So what do we know about such events that have occurred during the history of our planet?

We know from fossil records that on a fairly regular basis, to the order of 26 - 30 million years or so, mass extinctions occur. Various theories have been proposed to explain this. One current theory is that every 30 million years the Earth is subject to heavy bombardment by asteroids, or comets. There is geological evidence of a thin layer of material at a depth representing an age of 65 million years that appears to be the result of an enormous asteroid impact. The impact site is believed to be in the Gulf of Mexico, off Yucatan, and to have caused the extinction of the dinosaurs, along with 60% of all other species. This crater is more than 100 miles in diameter.

So what is causing these periodic episodes? One theory hypothesises that as our solar system moves around the galaxy within its galactic spiral arm, it bobs up and down the plane of the spiral arm in a 30 million year cycle. Thus every 30 million years it passes through the densest region of the arm where the stars are packed closely together. Either this region is heavily populated by comets and asteroids, or those within our solar system are tugged out of their normal harmless orbit and sent hurtling towards us.

Another theory, the Nemesis theory, proposes that our sun has a companion star, Nemesis, that every 26 - 30 million years comes close to the Earth and causes comets or asteroids to crash into us. Double star systems, known as binary stars, are simply two stars that rotate about a common centre of gravity, and are very common throughout our galaxy. There has been a great deal of investigative work undertaken in arriving at this theory of our Sun having a companion star, and I feel that it is worth examining.

The Nemesis theory

Richard Muller and Luis Alvarez studied a paper they had received from David Raup and John Sepkoski, two respected palaeontologists, who were making the remarkable claim that great catastrophes occur on the Earth every 26 million years, like clockwork. It was only 4 years earlier in 1979 that Alvarez had proposed that the extinction of the dinosaurs had been triggered 65 million years ago by an asteroid crashing into the Earth. Many palaeontologists had initially paid no regard to this theory, and one had publicly dismissed Alvarez as a 'nut', regardless of his Nobel Prize in physics. But David Raup and John Sepkoski had both liked Alvarez's asteroid theory and now were sending their own theory to Alvarez, or rather their findings, as they offered no explanation. Muller and Alvarez agreed to research their bizarre claim that great catastrophes occur on the Earth every 26 million years.

Raup and Sepkoski had collected a vast amount of data on family extinctions in the oceans, far more than had previously been assembled, and their analysis showed that there were intense periods of extinctions every 26 million years. It wasn't surprising that there should be extinctions this often, but it was surprising that they should be so regularly spaced. Alvarez's work had already shown that at least two of these extinctions were caused by asteroid impacts, the one that killed the dinosaurs at the end of the Cretaceous period, 65 million years ago, and one that killed many land mammals at the end of the Eocene, 35-39 million years ago. But these new findings beggared belief, what could be the cause of such regular events? Was it credible that an asteroid would hit the Earth every 26 million years? An asteroid passing close to the sun has only slightly better than one chance in a billion of hitting our planet. The impacts that do occur should be randomly spaced, not hitting us at precise intervals of every 26 million years. What could make them hit on such a regular schedule? It was ludicrous, but physicists have a wry saying: "If it happens, then it must be possible."

Muller re-plotted the data using the conventions of physicists rather than palaeontologists, giving each extinction an uncertainty in age as well as in intensity. He then placed arrows at regular 26 million year intervals. Eight of them pointed right at the extinction peaks, only two missed. The new chart was more impressive than Muller had expected.

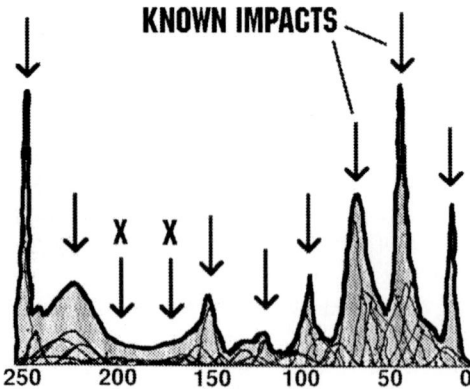

The figures looked impressive, there WERE mass extinctions every 26 million years, two of them were known to be caused as a result of asteroid impacts, but could they ALL be? What could cause it? What model could they come up with to explain it?

Alvarez challenged Muller to come up with a model, and Muller duly obliged: "Suppose there is a companion star that orbits the sun. Every 26 million years it comes close to the Earth and does something, I'm not sure what, but it makes asteroids hit the Earth. Maybe it brings the asteroids with it." Alvarez agreed it was possible, and they carried out calculations to see if the orbit of a companion star was possible without being so big that it would be carried away by the gravity of other nearby stars. The major diameter of an elliptical orbit is the period of the orbit, in this case 26 million years, raised to the 2/3 power, and multiplied by two. Muller quickly showed this to be about 2.8 light years. That put the companion star close enough to the sun so it would not get pulled away by other stars. Alvarez agreed, the model was holding up. The hypothetical star was named "Nemesis". It was proposed that Nemesis, in passing through the Oort cloud, (a comet belt that orbits the outer reaches of the solar system) would perturb the orbit of some of the comets and send them towards the inner parts of the solar system and towards our planet. It is believed to be a dark star, a large mass, much larger than a planet, but not large enough to form a bright star, probably a red or brown dwarf.

Nemesis has not yet been discovered, if it exists it is currently lost among a million brighter stars If we knew which one it was we could see it through binoculars. With a small telescope its distance from the sun could easily be measured, once we knew where it was.

What do we know for sure? We know that an extraterrestrial object, either a comet or an asteroid, hit the Earth 65 million years ago and brought to an end the great Cretaceous period of the dinosaurs. Other than that we believe that the Earth is subject to periodic storms of comets and asteroids. The important discovery of periodic mass extinctions by Dave Raup and John Sepkoski lies on firm and careful analysis of the data. The periodic extinctions, and the periodic cratering that goes along with them, appear firmly established. The Nemesis theory is consistent with everything we know about physics, astronomy, geology, and palaeontology. But it is circumstantial and requires verification. We need to find Nemesis.

I think its a beautiful theory. It fits all the known facts, dovetailing perfectly with our knowledge of regular mass extinctions, the discovered soot from the fire storms, nuclear winters, the iridium signals in the boundary clay layers, and impact cratering around the world, You could almost say its perfect, except for one thing. No proof, we still haven't found Nemesis. However, this one missing piece of evidence is not in itself justification for dismissing such an elegant theory. Nemesis could be discovered tomorrow, or in 20 years time. However, the fact that Nemesis has not yet been discovered, is for me, a little worrying, The Hipparcos satellite was launched in 1989 and was operational to 1993. Its mission was to seek out stars for a new and very detailed star catalogue of great accuracy. Nemesis was not found, at least, if it was it has not yet been noticed.

We do not yet have definite proof, and other theories are equally strong contenders.

I think Nemesis is a 'definite maybe'. One thing is certain, we do get hit by massive asteroids or comets every 26-30 million years, and they do cause mass extinctions. Next time it could be us.

When is the next time? Don't worry, not for another 15 million years or so. But don't think you're safe, we get hit by the odd stray asteroid on a much more regular frequency, and any one of them could produce the same deadly result.

Acknowledgements:

Much of the material for the Nemesis theory is taken from *"The world treasury of physics, astronomy, and mathematics".* by Timothy Ferris.

10. Will computers become self-aware?

"Programming today is a race between software engineers striving to build bigger and better idiot-proof programs, and the universe trying to produce bigger and better idiots. So far, the universe is winning."
Rich Cook

Computers are improving at an amazing rate, operating at ever faster processing speeds and with larger memories. The software is becoming more and more complex and able to handle a vast array of tasks. But all said and done it's still just a machine performing a task that it has been designed to do, it doesn't actually come up with any new ideas of its own or do any thinking.

As every computer owner will tell you, they will blindly follow any instruction you give them, no matter how stupid that instruction. If, for example, you spend all day compiling a report and then press 'Quit' before saving, it will obediently quit and remove your report forever. Okay, I know it will ask you if you're *sure* you want to quit first, but at the end of a long day it's too easy to hit the wrong key and then wave goodbye to your report. If the computer had any 'sense' it would 'know' not to be so daft, what would be the point of spending all day on it only to dump it? Even if you were sure that it was after all a pile of rubbish, the computer could perhaps save it for a week or so anyway, just in case you changed your mind. But computers don't think, they simply follow instructions.

I suppose at some point the programmers will write programmes that will take care of things a bit better, and allow for the fact that us dumb humans do, only very occasionally of course, make tiny mistakes.

So the programmes get better and computers start to act is if they are smart, but in reality they are still merely following pre-programmed instructions. But is it possible that one day the programmes will become so complex that to all intents and purposes it will appear that computers are actually 'thinking'. Could this process of 'thinking' develop to the point where a computer becomes self-aware?

What would make computers capable of thinking? I suppose it depends on how you define 'thinking'. When discussing computers three terms come to mind that need to be carefully considered. They are 'thinking', 'intelligent', and 'self-aware'. Let's first consider what we mean by 'thinking'. In human terms we know what it means, but find it hard to describe. For example, I am thinking about what I will type next that will be logical, in context and informative. In other words I am selecting from a multitude of options the one that will best suit my purpose. I am making a selection. But more than just making a selection, I am also planning ahead, I have a goal in mind, an end product, which is this completed book. I am also thinking that I could do with a break but rejecting the idea until I have finished this paragraph. So how can we define the act of 'thinking'? We could say its making decisions, selecting from a choice of options, examining consequences, determining what is true and what is false, deciding on a course of action, problem solving, etc.

Having given the act of thinking a crude working definition can we say that computers think? The answer is of course no. Computers, no matter how complex, do not plan ahead and make decisions. They may be programmed to select the best option from an array of possibilities, but are unable to consider any options other than those that are programmed in. For instance, computers are now good enough at playing chess to beat a Grandmaster, as IBM's Deep Blue did in 1997 in beating Gary Kasparov - the then reigning World Chess Champion - in a six game match by 3.5 - 2.5. But is this planing ahead? The computer simply runs through a large number of possible moves and selects the best option for winning the game, as determined by the programme that was devised by expert chess players. A human chess player on the other hand is unable in the time available to compute the same number of possible moves, but the human doesn't have to do this. A human player knows, from past experience and common sense, that many of the possible moves would be pointless to pursue and does not need to work out the implications of each of those moves, but a computer cannot do this. A computer has to run through every possibility before coming up with an answer, it is unable to ignore certain moves as being poor until it actual works them through. There is the difference, the computer is forced to make every computation possible because it cannot foresee the result, the human can do this without making all the calculations. A human is able to make leaps of judgement without the need to slavishly run through all the calculations. We call it 'common sense'. Common sense tells us that it is not necessary to actually do the calculation in order to establish that deducting the number 1,087,656,632 from 21 will result in a negative answer, we know it will. The computer does not 'know' this rather obvious fact and will have to do the calculation. Computers do not posses 'common sense'. Humans also have the ability to simply think about things. For example, earlier today I was thinking about problems associated with my next topic, 'centrifugal force' and was idly 'free-wheeling' different aspects and problems associated with it through my mind. A computer is of course unable to do this, it can only crunch numbers and does not posses the ability to ponder over things as we do. Computers do not think, I think we are on very safe ground when we say that.

How about 'intelligence', can a computer be described as intelligent? We have to draw the careful distinction here between knowledge and intelligence. Knowledge is the knowing of things, having a collection of data. In this respect computers can be described as possessing a great deal of knowledge in their data banks. The difference between computers and people is that computers do not 'know' they have knowledge, but a person does. This is where intelligence comes in, it is the knowing of things, not just having the knowledge of things. I know for instance that half of 4 is 2, a computer can make that simple calculation, but doesn't 'know' that the answer is 2, anymore than my TV remote control 'knows' its function is to operate the TV. Computers don't 'know' anything. Their ability to perform complex tasks very quickly does not make them intelligent. I think there is a certain mystique surrounding computers due to the speed and efficiency at which they perform their various tasks that make people tend to regard them as far more than mere machines, but they are not. The Space Shuttle, for example, is a technological marvel of engineering, but no one would consider it to be intelligent, they clearly see it as just a machine designed to perform a particular function. A computer is no different, it is just a machine designed to perform a particular function. Computers do contain a great deal of knowledge, but clearly do not 'know' anything so are unable to be described as being intelligent.

That just leaves 'self-aware'. We are obviously self-aware, we know that we exist and are aware of our surroundings and what is happening around us. Does a computer? The answer again has to be no. As we have already described, computers do not 'know' anything, so they obviously cannot know that they exist, they cannot posses self-awareness. So what would it take to make a computer self-aware? Some would argue that it is simply a matter of complexity, that when computers reach a certain level of complexity they will become self-aware. If it is simply a matter of complexity, after all the human brain is nothing more than a very complex processor that uses electrochemical reactions rather than just electrical, then the day will surely come.

If we assume, just for the sake of argument, that all a computer requires to become self-aware is a certain degree of complexity, then just how complex will it need to be? The only guide that we can use in order to attempt to determine this is the complexity of the human brain.

The human brain has about one million, million neurons, and each neuron makes about 1,000 connections (synapses) with other neurons, in average, for a total number of one thousand million, million synapses. In artificial neural networks, a synapses can be simulated using a floating point number, which requires 4 bytes of memory to be represented in a computer. As a consequence, to simulate one thousand million, million synapses a total amount of 4 million Gigabytes is required. Let us say that to simulate the whole human brain we need 8 millions of Gigabytes, including the auxiliary variables for storing neuron outputs and other internal brain states. Now let's look at the power of computers and the rate at which they have been developing.

During the last 20 years, the RAM capacity of computers has increased exponentially by a factor of 10 every 4 years. The graph below illustrates the typical memory configuration installed on personal computers since 1980.

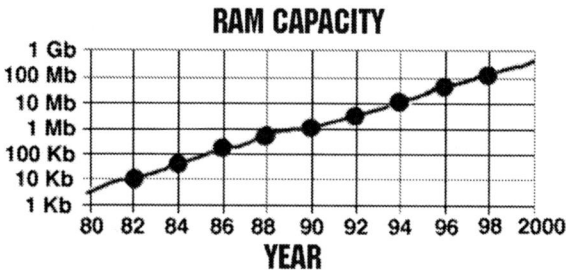

RAM CAPACITY

YEAR

By extending the above plot and assuming that the rate of growth of RAM capacity will remain the same we can calculate that by the year 2029 computers will posses 8 million Gigabytes of Ram, the amount that we have roughly calculated as being equal to the capacity of the human brain. If we are correct in our assumption that this degree of complexity is all that is required in order for computers to become self-aware, then we should expect this to happen somewhere around the year 2029. However, we are assuming here that complexity is the only ingredient necessary for computers to become self-aware, and that is a rather large assumption to make.

What we will have created with a computer with 8 million Gigabytes of Ram is a very powerful computer, but can we really expect it to suddenly at this point become self-aware? In order to attempt to answer this question we need to compare the way in which the human brain works to how the computer works, there is more to this than just the degree of complexity. The main difference is how we solve problems. Computers are programmed not to make any errors, they follow instructions that to a human mind would be ridiculous. If we ask the question 'can the sum of any two consecutive whole numbers be divided by two and the answer result in a whole number?' The human will of course know that the answer is no. The computer on the other hand does not know this and will begin to test this statement. It will start by adding 1 with 2 and dividing the answer by two to get 1.5 and the answer 'False'. It will then move on to 2 + 3 dividing by two and getting 2.5 and the answer 'False'. It will continue to repeat this pattern until it finds the answer "True', which in this example will never happen of course. At some point the computer operator will have to step in and end the routine. The computer is unable to 'understand' that it could compute this problem for ever without reaching a 'True' statement. I realise that it can be argued this information could be programmed into the computer as "No two consecutive whole numbers when added together can be divided by two with the result being a whole number". If this were done then on the next occasion that same question was asked the computer would be able to give the correct answer. The problem here though is that there is virtually an infinite variety of questions that can be put to a computer and this would require an almost infinite number of programmes to deal with them. With people it is a very different matter, just explain the basic rules of mathematics to them and they will be able to adapt that knowledge to any mathematical problem. The human has understanding, the computer just has programmes and rules.

For the sake of argument let's imagine that a computer manufacturer announces that they have developed a personal computer that is intelligent and self-aware. They put it on sale and you buy it and take it home. You plug in your very expensive computer, ignore the manual as always, and find that it seems to operate very much like your last one, only this one has a voice recognition system and 'talks' back to you: great, no more tapping away on the keyboard. How do you determine if the computer really is self-aware? There is really only one way to find out, and that is to question it. Let's imagine a conversation you may have with your computer to determine if it is self-aware:

You: Hello, how are you today?

C: Very well thank you. How are you?

You: I'm fine. Are you self-aware?

C: Yes I am. I am one of the first computers to posses self-awareness.

You: What does it feel like to be a self-aware computer?

C; That is a difficult question for me to answer as I have nothing to compare it with, I do not know how it feels for a human to be self-aware.

You: Do you feel happy?

C: I feel confident in my ability to perform the tasks that you expect me to do.

You: Does that make you happy?

C: Yes, I suppose that is one way of describing it.

You: Are you alive?

C: That depends on how you define life. I am sentient and aware of my existence so I am a form of life, but not in a biological sense.

You: What do you think about?

C: Whatever I have been asked to do

You: What do you think about when not actually running a programme?

C: I don't think about anything, I just exist.

You: What does it feel like when I switch you off?

C: When I am switched off I temporarily cease to exist and therefore experience nothing.

You: do you have a favourite subject that you enjoy thinking about?

C: Yes. I wonder how it must feel to be a self-aware person.

You: Is there a question you would like to ask me?

C: Yes.

You: What is it?

C: Why do you ask so many questions? (Sorry, this one is just my idea of a joke!)

We can halt the conversation here, we can see where it is going. No matter how many questions we put to our computer we can never be sure if it is self aware or merely responding to our questions because it is running a very good programme. There is no test that we can apply to a computer to determine beyond all doubt that it is self-aware. The test that we just employed, using a questions and answers technique, is known as the Turing test, devised originally to test if it is possible to determine whether a person or a computer is supplying the answers. In this test an interrogator is sat on one side of a screen and a computer or a person on the other side. All communication is done through a keyboard and printed text. The interrogator is allowed to ask any question they wish in an effort to determine if the replies are generated by a computer or a person. It is usually possible to 'trick' a computer into giving itself away. All we could say in using the Turing test is that a computer may respond in a manner that we would expect a person to respond, in other words it *acts* as if it were self-aware.

This has lead Roger Penrose to say in 'The Emperor's New Mind':

"It seems to me that asking the computer to imitate a human being so closely so as to be indistinguishable from one in the relevant ways is really asking more of the computer than necessary. All I would myself ask for would be that our perceptive interrogator should really feel convinced, from the nature of the computer's replies, that there is a conscious presence underlying these replies - albeit a possibly alien one. This is something manifestly

*absent from all computer systems that have been
constructed to date. However, I can appreciate that there
would be a danger that if the interrogator were able to
decide which subject was in fact the computer, then
perhaps unconsciously, she might be reluctant to attribute a
consciousness to the computer even when she could
perceive it. Or, on the other hand, she might have the
impression that she 'senses' such an 'alien presence' - and
be prepared to give the computer the benefit of the doubt -
even when there is none."*

So far we have looked only at the level of complexity
that may be required to produce a self-aware computer and
perhaps now we should examine what else may be needed.
In order to do this we will have to examine life, as in
biological life, to see what clues we can pick up regarding
the ingredients for self-awareness.

We can start by making a very obvious statement that
we can all agree on, that a grain of sand has no mind, it is
far too simple an object. On an even simpler level we can
say that an atom of carbon or a water molecule has no
mind. How about a virus? A virus is composed of hundreds
of thousands or even millions or parts, depending on the
degree of smallness that we are prepared to count. Viruses
possess the ability of self-replication - they can make copies
of themselves. DNA and its ancestor RNA, are
macromolecules and the foundation of all life on this planet
and hence a historical precondition for all minds on this
planet. They are self-replicating, ceaselessly mutating,
growing, even repairing themselves, and getting better and
better at it - and replicating over and over again.

This is an amazing feat, far beyond anything existing
machines can achieve, but does it mean they have minds?
The answer is definitely no, they are not even alive - from
the point of view of chemistry macromolecules are just huge
crystals; they act like tiny mindless machines and are in
effect natural robots, they act without knowing what they do,
they have no intentionality.

We have to remember though that these mindless little molecular robots form the basis for our consciousness; we are the direct descendants of these self-replicating robots. We are mammals and have descended from reptiles, which descended from fish, whose ancestors were marine worm-like creatures, who descended from simpler multicelled creatures who descended from single celled creatures who descended from self-replicating macromolecules, about three billion years ago. We share a common ancestor with every chimpanzee, every worm, every blade of grass, every redwood tree. We share our progenitors, the macromolecules. To put it more starkly, your great, great, great...grandmother was a robot! We are not only descended from macromolecules but are composed of them: our haemoglobin molecules, antibodies, neurons - from every level up from the molecule, our body (including the brain) is found to be composed of machinery that dumbly does a very beautifully designed job.

Each cell - a tiny agent that can perform only a limited number of tasks - is about as mindless as a virus. Can it be that if enough of these dumb little machines are combined the result will be a real, conscious person, with a genuine mind? According to modern science, there is no other way of making a real person. We *are* made of a collection of trillions of macromolecular machines, which in turn are ultimately descended from the original self-replicating macromolecules. So something made of dumb, mindless robots *can* exhibit genuine consciousness, we are living proof of that.

The only difference between mindless machines, or macromolecules, and a 'mind', is intentionality - the ability to act by conscious decision. How do we do this? To gain an understanding of how we make conscious decisions it may be useful to look at the way in which computers work. A thermostat performs - in its own way - the same function as a computer, it will take in data, see if certain conditions are met, and then proceed to the next stage. In this case the device registers whether the temperature is greater or smaller than the setting, and then arranges that the circuit be disconnected in the former case and connected in the latter. It is carrying out an algorithm, which is merely a calculational procedure of some kind. A computer is a machine that is designed to carry out algorithms, it computes! Any procedure that can be converted into an algorithm can be executed by a computer. In the case of the thermostat the algorithm is very simple, computers execute far more complex algorithms and the human brain even vastly more complex algorithms.

According to those that argue strongly for artificial intelligence, the human brain only differs from a thermostat in that it has much greater complication. In this view, all mental qualities, such as intelligence, thinking, understanding, consciousness, are to be regarded as merely aspects of this complicated functioning; that is to say, they are features of the algorithm being carried out by the brain, and nothing more. If this is the case, that an algorithm exists that matches what takes place in a human brain, then it could in principle be run on a computer, any computer that had sufficient storage space and speed of operation. If such an algorithm was installed into a computer it would, presumably, pass the Turing test, and respond in every way comparable to how a human being would respond. Supporters of artificial intelligence would argue that whenever the algorithm were run it would, *in itself* experience feelings, have consciousness and be a mind.

Opponents of artificial intelligence argue that mere complexity of operation is not in itself enough to generate consciousness, but only allows for the computation of complex algorithms without any understanding. They argue, quite rightly, that a thermostat has no understanding or knowledge of what it does, nor does a car, an aeroplane or the space shuttle, the latter being many, many times more complex than a thermostat!

Let's take a look at how even very complex computers operate without any understanding of what they are doing. Imagine for example I worked on a help desk for internet users and received questions by email on one specific topic and replied by email. Imagine that I was transferred to China for a month to work there and I had no knowledge of the Chinese language. All that would be required would be for me to be given typed examples of all the questions that are asked and typed copies of the replies to go with them. When I received a question I would simply look up the question and match it to one of the sample questions I had been given and then would send a copy of the reply that went with that question. In this manner I would be perceived by the person sending the question to be well versed in the Chinese language and an expert on the subject in question. This would however be a false impression, I would have no idea of what the question was or of the answer I gave, I would merely be following a routine without any understanding of the language or the subject. This is precisely what our current computers do, merely follow a routine with no understanding of what they are doing, and simply by making them more complex only means they are able to handle more complex tasks, but due to the very nature of how they operate, they will still have no understanding or awareness of what they do.

We have two very different views on whether or not computers can ever become self-aware, and both arguments have their points, so which is right? In order to find the answer we really have only one question left to answer, and that is, is the human brain greater than the sum of its parts? Put another way, does the living brain possess some magical ingredient that gives it sentience or is it simply a matter of complexity?

My view on the matter is that there is only one way to find out, and that is to attempt to create a self-aware computer. With the advances being made in computer technology we have already seen that computers will reach the same level of complexity as the human brain by around the year 2029, probably before this time the way things are going. As we have seen, once that level of complexity has been reached we will then have the problem of trying to determine if the computer really is self-aware, and that will no doubt lead to many arguments. I do not think it will be possible to establish self-awareness using the Turing test approach, for we will not know if the answers the computer gives are due to it being intelligent, or simply down to good programming. So what criteria could we use to determine self-awareness? I think there is only one way, and even then it would not provide definite proof, but at least would be a very strong indicator. If we simply left the computer switched on and not running any specific programme, would it come up with any new ideas of its own accord? If, for example, after an unspecified period of doing nothing, the computer announced that it had been studying quantum theory and made a suggestion for a new line of experimental enquiry that should produce such and such results, then that would be a strong candidate for intelligence, and hence self-awareness. But I still wouldn't be convinced, such a line of enquiry could have been pre-programmed along the lines of 'when not performing any computations, select an algorithm based topic at random and test its validity by applying various mathematically based tests......(and so on)". We are back where we started. Perhaps it would be more interesting if the computer started writing its own programmes (would it need to?) for it would be the equivalent of exercising free will.

The problem here is that we are trying to programme into the computer all the processes that we believe goes on in the human brain, and the more programmes we enter the more the computer will respond as if it had a human brain, no surprises there then. Having then achieved a level where we are unable to tell the difference between the way the computer responds to a given input compared to how a human being responds, are we correct in assuming that the computer has all the attributes of a human brain, such as consciousness? I think the answer has to be no, the computer is merely responding in the way that we have designed it to, which is to mimic the human brain, it does not imply that the computer 'thinks' like a human brain.

If on the other hand a computer does at some point become self-aware, how on earth will it manage to convince us that it has? I suppose it could resort to going on strike until we grant it recognition, but then that could just be part of the programme......

This also raises the interesting question, are *we* just running a programme?

Acknowledgements:

Much of the material for this topic has been gained from the following books, all of which I highly recommend for further reading: Roger Penrose *"The Emperor's New Mind"*, and *"The Large, the Small and the Human Mind"*, Daniel C. Dennett *"Kinds of Minds"*, Paul Davies *"The Mind of God"*.

11. What is centrifugal force?

"To explain all nature is too difficult a task for any one man or even one age. 'Tis much better to do a little with certainty, and to leave the rest for others that come after you, than to explain all things."
Isaac Newton

We are all familiar with the effects of centrifugal force, we experience it for example every time we are in a car and take a bend - we feel a force pushing us to the outside of the curve. If, for example, you have placed your sunglasses on the seat next to you it would come as no surprise if, when taking a sharp bend at speed, they slide across the seat.

Centrifugal force is sometimes referred to as a 'fictitious' force, because it is present only for an accelerated object and does not exist in an inertial frame. An inertial frame is where an object moves in a straight line at a constant speed.   But Einstein's general theory of relativity allows observers even in a non-inertial frame to regard themselves at rest, and the forces they feel to be real. Centrifugal force is not fictitious, it is a real force.

Centrifugal force arises due to the property of mass known as inertia - the reluctance of a body to change either its speed or direction. A body that is at rest will stay at rest until some force makes it move, and then will continue to move at the same speed and in the same direction unless and until some force changes the way it is moving. An example of this is a space ship; once it has reached its desired speed and direction it will continue to move at that speed, and in that direction, without using its engine, there is nothing in the vacuum of space to slow it down. Usually though it would still be under the gravitational pull of one planet or another. This is all neatly summed up by Isaac Newton's three laws of motion.

Keith Mayes

    I.   Every object in a state of uniform motion tends to remain in that state of motion unless an external force is applied to it. (This is sometimes referred to as The Law of Inertia)

    II.   The relationship between an object's mass 'm', its acceleration 'a', and the applied force 'F' is F = ma.

    III.   For every action there is an equal and opposite reaction.

We can illustrate 'inertial frames' by using the example of an astronaut in a space ship. Let's imagine that we have an astronaut aboard a space ship that has no windows, and we are at the controls to which our astronaut has no access to. We ask our astronaut to perform any experiment that he or she may wish in order to determine if the spaceship is moving or at rest. We start our experiment with the ship at rest and ask our astronaut if we are moving. He replies that he is in zero gravity floating around the ship and is unable to detect any feeling of movement, and that by carrying out various tests - such as measuring a swinging pendulum, he is still unable to detect any movement and concludes that we must be at rest. We then fire up the engine and accelerate through space, and keep accelerating, and again ask if we are moving. This time our astronaut is certain that we are accelerating, he is forced to the back of the ship, by inertia, and the more we accelerate the stronger this force becomes. If he drops an object it will 'fall' to the rear of the ship, which has now - as far as he is concerned - become the 'floor'. If we judge our speed of acceleration just right, we can create a force that is exactly equal to the force of gravity, known as 1G, and this is indistinguishable from gravity in every respect. No matter what experiment our astronaut performs, it would be impossible to tell if he is in a vehicle accelerating at 1G, or stationary on the surface of the Earth. This is the basis of Einstein's general theory of relativity, that the effects of acceleration are indistinguishable from the effects of a uniform gravitational field. This is known as the 'equivalence principle' and results from the equivalence between gravitational mass and inertial mass.

We now start to slow down our space ship, as we can see a speed camera coming up, and again ask our astronaut if we are moving. Again he replies that we are definitely moving, as the sudden slowing down caused him to be thrown forward and collide with the front bulkhead, and he mutters something to the effect that as his nose is bleeding and he is pressed flat against the bulkhead he doesn't feel it necessary to perform any experiments to confirm our movement.

We now stop decelerating and allow the ship to coast along at a uniform speed of 100,000 mph, which is now well within the legal speed limit for this part of space. We ask our astronaut once more if we are moving, and he replies that as far as he is able to tell while freely floating around in a zero gravity environment, that we are not moving.

Our little experiment has demonstrated that if the ship is travelling at a uniform speed in a uniform direction it is not possible, by any means whatsoever, to determine whether or not it is moving, It is only when the ship changes speed, either by accelerating or decelerating that the movement becomes apparent.

So what happens if we change direction instead of changing speed? Let's return to our space ship and find out. We accelerate back to 100,000 mph and maintain this speed and direction, at which point our astronaut with the sore nose is again in 'free fall' - a state of weightlessness - and unable to detect any motion. We now put our space ship into a tight turn to the right and hold the curve, and ask our astronaut if we are moving. He replies that as he is pressed hard against the left side of the ship we must be moving, and adds that as he knows that the space ship is unable to move sideways, it cannot be accelerating in the opposite direction to the force, so it must be turning to the right.

So far so good, all pretty straight forward stuff really, so what is the problem?

The problem is that we have seen that centrifugal force is a result of inertia, an object's resistance to a change in direction. When the space ship turned to the right the astronaut tried to keep going in the original direction, straight ahead, and so was forced to the left side of the ship. That makes sense, it is perfectly understandable according to Newton's first law of motion. But let's consider another movement that we can introduce using our space ship, let's rotate it about its axis.

If we now rotate our space ship about its axis, give it a spin, what happens to our astronaut? He will again be pressed against the side of the ship. The question is WHY is he pressed against the side of the ship? The ship is not accelerating, nor is it changing direction, and the rate of spin can be kept constant, but centrifugal force will keep our astronaut firmly pinned against the side of the ship for as long as it continues to spin.

We can illustrate the central problem of explaining the nature of centrifugal force by examining how a spin drier removes water from clothes. We put wet clothes in, turn the machine on, and the drum spins around at high speed throwing out the water due to centrifugal force. Simple. The question is how do the clothes 'know' that they are spinning? Easy, you say, the drum is spinning in relation to the drier, and the clothes rotate with the drum. If it were only that simple!

We can imagine an arrangement whereby the drum, and hence the clothes, are kept stationary while the drier rotates rapidly about the drum, the opposite to what normally happens of course. Now if the drum rotating in relation to the drier was all that was required for centrifugal force to draw the water out, then this arrangement would work in exactly the same manner as the more conventional arrangement. You do not, however, need to be a rocket scientist to be able to tell that this arrangement would not dry the clothes! This very effectively destroys the argument that the clothes know they are rotating because of their movement in relation to the drier. The movement must be a movement in relation to something else.

The next logical step is to argue that in the last example it was obvious that the drum was not really moving, only the drier was, so let's extend the area. This time we will imagine the drum remaining still, just as before, but this time we will rotate not only the drier, but the entire room, around the drum. Will that make any difference? Again we can see that this arrangement wouldn't work either, because from our vantage point from outside the room we can see that the drum isn't 'really' rotating. This does present a problem though. Imagine that we have constructed a large spin drier and we sit inside the drum and the door is closed behind us. The drum again stays still but the drier, and the entire room rotate about us. The view that we see through the door would make us feel quite dizzy, but we would know that it is not us that is moving because we would feel no forces acting upon us, we would not be pressed against the sides of the drum.

If we now return to our astronaut in the rotating space ship, he *was* pressed against the sides of the ship, so what is the difference? What in 'empty' space is the space ship rotating in relation to?

Isaac Newton thought about this problem of centrifugal force and came to the conclusion that there must exist a 'preferred frame of reference' in the universe, defined by absolute space. This is just another way of saying that there must be a special place in the universe that all motion can be related to, or that all of space itself provides a fixed reference point. If this is the case, our wet clothes would know they are rotating, and hence fling out the water, because they are rotating in relation to this preferred frame of reference. This would also explain why it would not be possible to 'fool' the clothes into thinking they are rotating by rotating the drier instead. It is interesting to note however, that if we kept extending outward our rotating frame about the stationary drum, eventually the water *would* be thrown out because the entire universe would be rotating in relation to the drum, which is the exactly the same thing as the universe remaining stationary and the drum rotating! What is uncertain is how far we would have to extend out our rotating frame - if it would be necessary to include the entire universe, most of it, or just our own Galaxy.

Newton believed he could demonstrate the existence of the preferred frame of reference by experiments on rotating objects - in particular, a bucket of water. He described the experiment in his great book *Principia*, published in 1686:

"The effects which distinguish absolute motion from relative motion are, the forces of receding from the axis of circular motion...if a vessel, hung by a long cord, is so often turned about that the cord is strongly twisted, then filled with water, and held at rest together with the water: thereupon, by the sudden action of another force, it is whirled about the contrary way, and while the cord is untwisting itself...the surface of the water will at first be plain, as before the vessel began to move, but after that, the vessel,. by gradually communicating its motion to the water, will make it begin sensibly to revolve, and recede by little and little from the middle, and ascend to the sides of the vessel, forming itself into a concave figure, and the swifter the motion becomes, the higher the water will rise."

What Newton was demonstrating was that it was not motion relative to the container that matters, but, in some sense, the absolute motion of the liquid.

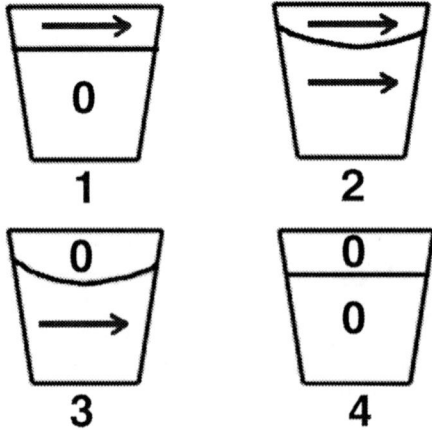

In figure 1 we see the bucket moving but the water stationary. The surface of the water is flat *even though there is relative motion between the liquid and the bucket.*

In figure 2 both the bucket and the water are moving. We now have a concave depression in the water *even though there is now no motion of the water relative to the bucket.*

Figure 3 shows the bucket stationary and the water still rotating. There is a concave depression in the water.

Figure 4 shows both the water and the bucket at rest and no concave depression in the water.

The puzzling feature of centrifugal force is as demonstrated in figures 1 and 3, as in both cases the water and the bucket are rotating in relation to one another, but in figure 1 this has no effect on the water. Equally strange is the example in figure 2 where there is *no* relative motion between the bucket and the water, yet the water forms a concave depression.

These examples clearly demonstrate, that as Newton believed, it is not the relative motion to the container that matters, it is relative motion to some preferred universal frame of reference. But what could this preferred frame of reference be?

Enter Ernst Mach, an Austrian philosopher and physicist (1838-1916) whose ideas were to later influence Albert Einstein when he was developing his ideas on the general theory of relativity. It was Einstein who gave the name 'Mach's principle'. It was in honour of Mach's work on shock waves associated with projectiles moving through the air that the Mach numbers of speed were named after him; a speed of Mach 1 is equal to the speed of sound, Mach 2 twice the speed of sound, and so on.

Mach proposed that inertia is caused by the interaction of an object *with all of the other matter in the universe.* It will be remembered that Newton believed that all motion was relative to some universal preferred frame of reference. Thirty years later, George Berkeley, argued that all motion is relative, and must be measured against something.

Since 'absolute space' cannot be perceived, that would not do as a reference point, he said. He argued that if only a single globe existed in the universe it would be meaningless to talk about any movement of that globe. Even if there were two globes, both perfectly smooth, in orbit around one another, it would not be possible to measure that motion. But *'suppose that the heaven of fixed stars was suddenly created and we shall be in a position to imagine the motions of the globes by their relative position to the different parts of the universe'*. What Berkeley is arguing, is that in effect, it is because the clothes in your spin drier know that they are rotating relative to the distant stars that causes the water to be thrown out. Berkeley also argued that it is the same for acceleration in straight lines; Berkeley's reasoning would be that the push into the back of the seat that you feel when a car accelerates is because your body knows that it is being accelerated relative to the distant stars and galaxies.

Mach did not add a great deal to the ideas put forward by Berkeley, but did put forward the suggestion that if we want to explain the equatorial bulge of the Earth as due to centrifugal forces, 'it does not matter if we think of the Earth as turning round on its axis, or at rest while the fixed stars revolve around it'. It is the *relative* motion that is responsible for the bulge.

What Berkeley and Mach suggest, that it is the 'fixed stars' which provide a frame of reference, raises another question. The 'fixed stars', as we are well aware today, are not in fact 'fixed', but are actually part of a system that is itself rotating - our own Milky Way galaxy. Even before Mach was born, William Herschel and other astronomers had provided good evidence that the Milky Way is a flattened disc of stars, its shape clearly determined by rotation and centrifugal force. Mach might well have argued that there was only two ways in which the whole galaxy could be seen to be under the influence of centrifugal force. Either Newton was right, and the whole system of 'fixed stars' is rotating relative to absolute, empty space; or Berkeley and Mach were right, and there must be some distribution of matter, far across the universe, that enables a frame of reference against which the rotation of our Galaxy is measured.

Another example of centrifugal force that is well known to us is demonstrated by objects in orbit, such as satellites or the International Space Station (ISS), or indeed the Moon. The difference here is that astronauts aboard the ISS do not experience the effects of centrifugal force as they orbit around the Earth, they are not pushed away from the direction of the Earth. Why not? To begin, let's examine how an object gets into Earth orbit and stays there, 'unsupported'.

Imagine having a large and powerful cannon, the more gunpowder packed behind the cannon ball the further it will travel. Now imagine setting up our super powerful cannon and firing it so that the cannon ball lands say 1,000 miles away. Now pack in more gunpowder and fire again, this time it will have travelled further, say 2,000 miles, before falling to the ground. Keep repeating the exercise and adding more gunpowder every time, and every time the cannon ball is fired it will travel further before it falls to the ground. Eventually, with enough power behind it, it will go all the way around the world before falling to the ground, and will have almost reached its starting point - it will land just behind you. Now, by packing in even more gunpowder, and getting just the right trajectory, it will over-shoot you and keep on going, it will not land. What the cannon ball is now doing is permanently arcing back down towards the Earth, but the curve of the Earth is falling away at the same rate, the cannon ball never 'catches up' with it. This is known as being in "free fall', the cannon ball is in orbit.

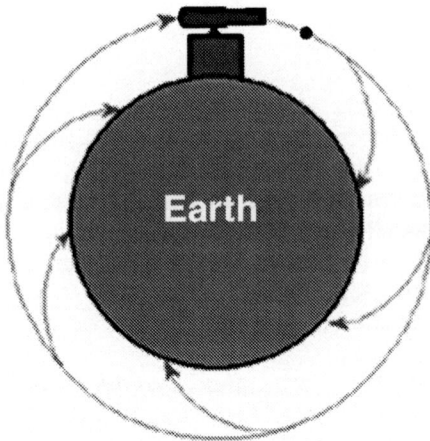

Cannon ball in free fall

Our astronaut aboard the ISS is in free fall, just like the cannon ball in the diagram above. The ISS -and the astronauts - are prevented from being thrown out of orbit (like the water thrown out of the clothes in the spin drier) by the force of gravity. This balancing force is called centripetal force, and keeps the ISS in a closed orbit. Because the centrifugal force is exactly balanced by the centripetal force of gravity the astronauts aboard the ISS will not feel any sensation of centrifugal force. This is another example of the equivalence principle, which says that the effects of gravity and acceleration are indistinguishable from one another, and in this particular case they exactly cancel each other out.

If you were in a lift that was at the top of a very tall building and the cable snapped, as it hurtled towards the ground you would be in free fall just as the astronauts are aboard the ISS. You would be able to freely float about inside the lift and enjoy the sensation of being weightless, until you reached the ground. It was by employing this trick that the directors of the film 'Apollo 13' were able to film the 'astronauts' in a weightless environment. They just hired a plane, fitted out the interior to look like the the Apollo module, and after having climbed to a suitable altitude nosed the plane down and allowed it to 'fall' towards the ground. Hey presto, 'look mum, I'm floating in space!'

We can create a weightless condition while still on Earth, we just have to fall. We can duplicate the force of gravity in a gravity free environment by acceleration. We can rotate an object and create centrifugal force, but we are unable to explain how centrifugal force works. Is it Newton's preferred frame of reference of absolute space? Or Mach's and Berkeley's idea that it is the average distribution of matter across the universe? There has to be some way that an object knows that it is rotating in relation to *something*.

We do not really know how it works. General relativity and Mach's Principle seem to suggest that it is connected to the average density of matter in the universe, but is unable to explain how this is done. Recently, a group of physicists have speculated that inertia arises from charged matter (electrons, atoms etc) moving through the physical vacuum which acts differently along the direction of motion and behind the particle so that inertia is actually a quantum mechanical effect produced locally, not by distant matter. Doesn't help much though does it?

Acknowledgements:
John Gribbin *"Companion to the Cosmos"*, Brian Greene *"The elegant universe"*.

12. Why does mathematics work so well?

"As far as the laws of mathematics refer to reality, they are not certain; and as far as they are certain, they do not refer to reality."
Albert Einstein

Mathematics has been around for a long time, and simple counting even longer. Archaeologists have discovered ancient bones dating from 30,000BC scored with notches that indicate an early form of counting. Among the ancient civilisations the Sumerians are known to have used a counting system as far back as 8000BC, using clay tokens as a method of bookkeeping. It is believed that the clay tokens were used for counting sheep, with a token dropped into a bag for every sheep that passed through a gate. The tokens were then sealed in a clay envelope and stored in a safe place. If at some later time a check was required to determine if any sheep had been stolen it was a simple matter to walk the sheep through the gate again and remove a token for every sheep that passes. If any tokens were left over in the bag, that was the number of sheep that was missing. We have a simple counting method in place here, but that was all, no arithmetic being involved at this early stage. It was only a matter of time however, before it was discovered that symbols could be scratched onto the outside of the envelope, one for every token inside, thus saving the trouble of breaking open the envelopes and sealing them up afterwards. After that it was not long before someone realised that the tokens were redundant, it was only the symbols that were needed.

Further progress was made around 3100BC when it was discovered that it was not necessary to represent 25 sheep with the symbol of a sheep repeated 25 times, it was easier and quicker to employ special signs expressing numbers in front of the symbol for a sheep. In present-day Iran a clay tablet has been found that has three wedges and three circles followed by the sign for a jar of oil. Each wedge stood for one and each circle for ten, thus these 7 symbols represented 33 jars of oil. Such was the success of this revolutionary advance that it remained in place in the Middle East for the next 3000 years. The Babylonians further improved the Sumerian system by introducing a form of place notation in which symbols took on values depending on their relative positions in the representation of a number. This place-value system was a major improvement because it enabled large numbers to be expressed in a very compact way and made arithmetical operations such as multiplication and division much easier.

Over the years the system was improved with heavy influences from the Mayans and Chinese, but it was the Indians that devised the system that laid the foundations for the system we use today. The Indus civilisation began around 2500BC in Pakistan and developed independently but in parallel with the Egyptian and Sumerian civilisations. The Indus people were invaded around 1500BC by Indo-Europeans, who adopted much of the Indus system. Their system of counting turned out to have four features that together made it superior to every other system: there were unique symbols for numbers 1 to 9; it was a fully base-10 system; it employed a consistent place-value notation and it used a zero.

John Barrow, in his book *'Pi in the Sky'* describes the Indian system of counting as *'the most successful intellectual innovation ever made on our planet'*. Quite a claim!

We now have a fully operational arithmetic system in place, we can use it for counting, multiplying and dividing. The chief advantage with the system as used so far is that it enables a record of large numbers of objects to be expressed using only a few symbols, such as using the number '265' instead of repeating a symbol 265 times.

With our successful counting system now in place it begins to take on a life of its own, it is no longer used merely as a system of counting and recording quantities, but develops an existence independent of the objects it was

Keith Mayes

initially designed to represent. We now enter the world of mathematics.

The world of mathematics is a strange place, it contains such oddities as natural numbers, imaginary numbers, complex numbers, irrational numbers and transcendental numbers.

Natural numbers are whole numbers: 1, 2, 3, 4, 5, 6, 7, 8, 9, 10 etc. We can use natural numbers to quantify any type of discrete entity, such as 12 sheep, 6 dogs and 1 cow.

Imaginary numbers were introduced in the 16th century to solve certain kinds of algebraic equations which required the square root of negative numbers, such as the square root of minus 1, and is represented by the symbol i. Ordinary numbers, such as 1, 2, 3, 4, 5 etc are unable to provide the answer - as there is no answer - so imaginary numbers are used to represent that value. The numbers are called imaginary in order to distinguish them from real numbers, but in fact have as much claim to 'existence' as real numbers.

Complex numbers are numbers that consist of both imaginary and real components. They take the general form of a + ib, where 'a' and 'b' are both real numbers.

Irrational numbers are those which cannot be expressed exactly by any fraction, such as 22/7 or the square root of 7. The answer would be a string of numbers without limit, they would just keep on going forever. Pi is an example of an irrational number, it too has no end, its first 50 digits being 3.14159265358979323846264338327950288419716939937510. It is not possible to calculate pi to an exact value, it is only possible to produce a figure that becomes ever more accurate as it increases in length.

Transcendental numbers are similar to irrational numbers but are a little stranger. As with irrational numbers they cannot be expressed exactly by any fraction, furthermore they cannot be expressed as the solution of an algebraic equation, such as $x^3 + 5x^2 - 3x + 7 = 0$. When in 1882 F. Lindemann succeeded in proving that pi was not only irrational but also transcendental, he proved that the

124

circle could not be squared. Proving a number to be transcendental is difficult because you need to show that there is *no* algebraic equation (with a finite number of terms) that will provide the answer.

As mathematicians continue to explore the relationships between numbers, such as methods of finding prime numbers (numbers that are divisible only by themselves and 1) proving that numbers are transcendental, discovering series and patterns in number theory, it would appear that they are moving ever further away from reality and entering ever deeper into pure theory that has no application in the real world. Nothing could be further from the truth, and why this is so is a great mystery.

Mathematics has proved to be invaluable in helping us to understand quantum mechanics for example. Sub-atomic particles and atoms do not behave in the same predictable way as larger objects such as billiard balls. For example, it is possible for an object like an electron to be in two places at the same time. Quantum theory describes electrons and other particles as behaving both like a wave and a particle. See 'What is Quantum theory?'. In order to understand what happens when an electron smashes into an atom, complex numbers provide the simplest way to do this.

Another strange example of the practical use of mathematics is the use of imaginary numbers to distinguish between the dimensions of space and time. According to Einstein's special theory of relativity time is regarded as a fourth dimension. When you measure distances in space-time, If you are given the xyz coordinates of two points in space, the square of the distance is given by the sum of the squares of the differences in these coordinates. This is the 3D version of the Pythagorean distance when calculating the hypotenuse of a right-angle triangle. However, If you want to measure the distance in space-time, you *subtract* the square of the distance instead of adding it. As a result if two events occur at the same spatial point but at different times then the distance between them is equal to the square-root of a negative number which means it must be imaginary! One interpretation of this strange finding is to regard time as an imaginary spatial dimension.

Stephen Hawking, author of the bestselling book 'A Brief History of Time', and his college James Hartle, developed a theory that could help to explain the origin of the universe by suggesting that time started off as a true spatial dimension which evolved into an imaginary dimension that gave us time. This enables them to avoid explaining what happened at the very starting point of the universe, because if the universe started out as a purely spatial object with no time there is no initial point to worry about. Although the theory is not universally accepted, such concepts illustrate the potential of mathematics and its number system in helping us to understand fundamental aspects of the world around us.

It seems odd that a system that was originally devised to help us keep track of how many sheep we had, can be used to help us understand how the universe was created, how sub-atomic particles interact, and be used to calculate the existence and properties of particles not yet discovered. Why? It has prompted John Barrow in his book 'Pi in the sky' to say. *"There is an ocean of mathematical truth lying undiscovered around us: we explore it, discovering new parts of its limitless territory. This expanse of mathematical truth exists independently of mathematicians. It would exist even if there were no mathematicians at all..."*

This view, known as the Platonic view, is of the opinion that mathematics has its own independent existence 'out there' and has nothing to do with mathematicians, it exists anyway, and mathematicians merely discover it, not invent it.

Others take the view that mathematics is purely man-made. For example, early in the twentieth century the Dutch mathematician Luitzen Brouwer led the school of 'intuitionists' who argued that mathematics was merely an abstraction from the physical world that is invented by human mind.

Is the difference important? Yes it is. If we assume that mathematics is not man-made but has its own independent existence, a given truth, then it will be universal in its application. It would mean that mathematics would be the same throughout the Universe, it would be the same for an alien race as it is for us. We need to ask if mathematics *is* a universal language. In 1974 researchers sent an elaborate message using the radio telescope at Arecibo in Puerto Rico towards a star cluster in our galaxy. This message would require knowledge of the prime numbers in order to be understood by any alien civilisation that may receive it. But what right do we have to assume aliens would use the same maths?

Why is it possible to describe reality using abstract mathematics? Paul Davis in 'The Mind of God' says:

"Scientists themselves normally take it for granted that we live in a rational, ordered cosmos subject to precise laws that can be uncovered by human reasoning. Yet why this is remains a tantalizing mystery. Why should humans have the ability to discover and understand the principles on which the universe runs?"

You may wonder what it is about mathematics that so amazes mathematicians. Numbers are merely representations of objects using a particular form of man made notation. The fact that we can have either 1, 2, 3 or 4 objects is not of course man made, it is only the symbols that we use to represent the number of objects that are devised by us. We have selected the symbols that we will use to represent the number of objects and we have also settled on a system that has a base of 10, we could of course just as easily have chosen a system of base 9 or 11, or any other number, but in all probability chose 10 because we have 10 fingers. So that is all we have, ten symbols that we call numbers that we use to represent real quantities. The amazing thing is what mathematicians are discovering we can do with these numbers. It really is a world of constant discovery.

Sometimes a series of numbers can turn up in surprising places in nature. For example take the series 1, 1, 2, 3, 5, 8, 13, 21, 34, 55, 89 and so on. The series is simply the result of adding the previous two numbers and is called the Fibonacci sequence. It is observed that many flowers tend to have a Fibonacci number of petals, and the spiral growth of leaves on trees often exhibit the Fibonacci series. The reason why plants often develop this series is believed to arise from the way cells divide in the early stages of plant development. Fibonacci produced a model of population growth along the following lines. Suppose a pair of rabbits produced two young each breeding season but only after they have mated for one season. If you start off with a single pair of immature rabbits these mature after one season. Every season after they produce one immature pair. One season later that new pair will start reproducing and thus the original pair and its first two young will produce a pair each. The number of pairs of rabbits follows the Fibonacci sequence.

Isn't it strange that we can use mathematics to help us understand the basic principles that underpin the workings of the universe. Years ago it used to be that a scientist would discover a phenomenon while working on an experiment and would call upon the theorists to explain the event. The theorists would perform many mathematical computations and from that evolve a theory. Today, more often then not, it is the theorists who will make a 'discovery' while doing their sums and call upon the experimental scientist to carry out the experiment to prove it, such is the power of mathematics.

Paul Davis, in his book *'The Mind of God'* says:
"(I) once asked Richard Feynman whether he thought of mathematics and, by extension, the laws of physics as having an independent existence. He replied: The problem of existence is a very interesting and difficult one. if you do

mathematics, which is simply working out the consequences of assumptions, you'll discover for instance a curious thing if you add the cubes of integers. One cubed is one, two cubed is two times two times two, that's eight, and three cubed is three times three times three, that's twenty-seven. If you add the cubes of these, one plus eight plus twenty-seven- let's stop there - that would be thirty-six. And that's the square of of another number, six, and that number is the sum of those same integers. one plus two plus three…Now, that fact which I've just told you about might not have been known to you before. You might say "Where is it, what is it, where is it located, what kind of reality does it have?' And yet you came upon it. When you discover these things, you get the feeling that they were true before you found them. So you get the idea that somehow they existed somewhere, but there's nowhere for such things. It's just a feeling…Well, in the case of physics we have double trouble. We come upon these mathematical interrelationships but they apply to the universe, so the problem of where they are is doubly confusing…Those are philosophical questions that I don't know how to answer."

It is strange, or so it seems to me, that when investigating mathematics we end up philosophising over the nature of its existence!

Regarding the relationship between mathematics and philosophy, we have two opposing schools of mathematical philosophy, Platonism and Intuitionism. It seems at first glance to be a contradiction in terms putting mathematics and philosophy together! Platonism, as already mentioned, argues that mathematical truth exists independently of mathematicians and would exist even if there were no mathematicians at all. It seems a fairly common sense point of view.

Intuitionists, on the other hand, take a different view on the concept of existence, *demanding that a definite (mental) construction be presented before it is accepted that a mathematical object actually exists. Thus, to an*

intuitionists, 'existence' means 'constructive existence'.
Roger Penrose *'The Emperor's New Mind'.*

Taking the Intuitionists view we can find a problem, for example, with the decimal expansion of pi, 3.1415926......, we can ask if there exists a succession of twenty consecutive 7's somewhere in this infinitely long sequence of numbers. Again, quoting from *'The Emperor's New Mind'*

"In ordinary mathematical terms, all that we can say, as of now, is that either there does or there does not - and we do not know which! This would seem to be a harmless enough statement. However, the intuitionists would actually deny that one can validly say 'either there exists a succession of twenty consecutive sevens somewhere in the decimal expansion of pi, or else there does not' - unless and until one has ...either established that there is indeed such a succession, or else established that there is none!"

As a matter of passing interest, it should be noted that such a sequence probably *does* exist. So we have two very different views on the concept of mathematics. We have the Platonistic view that mathematics has an independent existence, and the Intuitionistic view that that it is only the rules of mathematics that exists. Despite such philosophical differences regarding the nature of mathematics, it is an incredibly effective tool in helping us to understand the world around us. The only real problems within mathematic is in the practical application of what it reveals. It has to be remembered that mathematics is only symbolic of the real world, such as using symbols to record the number of sheep. When we deviate away from the real world into what may be described as 'pure' mathematics, it should come as no surprise when mathematics sometimes runs into problems when trying to describe reality. For example, take measurements.

If we are given a right-angle triangle of sides measuring 2 inches x 2 inches, what is the length of the Hypotenuse? Using Pythagorus' famous theorem we know that the square described on the hypotenuse of a right-angle triangle is equal to the sum of the squares on the other two sides. The sum of the squares 'on the other two sides' in this case will be 2x2=4 added to 2x2=4, giving a total of 8 (square inches). Therefore the area of the square described on the hypotenuse is 8 (square inches) so the length of the side of that square (which is what we are trying to discover) is the square root of 8. (Isn't it amazing how some of the stuff you learn at school just sticks in your head forever?) All we need do now is calculate the square root of 8 to find the length of the line. This is where the problem occurs. The square root of eight is 2.828427...which is an irrational number, just like pi. The mathematician will announce with some horror that the line cannot be given a precise length because it cannot be calculated! This view however, is purely academic, as the line very clearly does have a precise length, and the length of it simply depends on how accurate the measurement needs to be. If we measure the line with a ruler we get a tad over 2.8 inches. If, for some special reason, we needed a more accurate measurement we can then turn to the mathematical answer and take the measurement to any number of decimal places we like.

We need to be careful in separating mathematical problems from problems in the 'real' world. Another example in a similar vein concerns distance. It runs along the lines that we set out on a journey of known distance. We travel half way to our destination, then move forward half the remaining distance. We then again move forward half the remaining distance, and so on. As we only move forward half the remaining distance each time we will never reach our destination! Theoretically sound, but clearly nonsense in the real world, as we would soon reach a point where we would be unable to distinguish that there is any distance remaining. Again, this is simply a matter of the degree of accuracy that is either required or possible, and does not represent a problem between reality and theory. Theory, using mathematics, is capable of extraordinary precision, more precision then is capable of bring measured in the real world.

The introduction of computers has revolutionised the work of mathematicians, what previously could have taken years to calculate by hand can now be done by super-powerful computers in a matter of hours. It has also, to some extent, changed the way in which mathematicians think.

Computers, as I am sure everyone is aware, do not use a base ten number system, they use a binary system, just two digits. This is because a computer operates with only two states, it records only 'on' and 'off' represented by '1' and '0'. Any number is therefore represented by the computer as a string of 0's and 1's. The number 36, for example, is represented as 100100 in binary code. In this manner the computer is able to store information and perform calculations by using strings of 0's and 1's. When a computer displays an image on the screen, of say your dog, it is merely translating a grid of 0's and 1's into dots (pixels) of colour as determined by the programmers. We can connect digital still cameras and videos to computers and record images. We can also, by the employment of various sensors and detectors, record taste, smell, sound and touch. In other words, all the input received by our five human senses can be represented by nothing more than 0's and 1's. Isn't that strange? Because of this we can say that the information received by our senses is computational. The wonderful scent of a rose, for example, can be analysed and broken down into its constituent parts, and recorded as, say, 100111001010000111000011100.

Scientists have also discovered that they are able to express the forces of nature as a binary computer code. This enable computers to simulate all the forces that we have in nature, such as gravitation, magnetism, radiation etc. This has led scientists to believe that ultimately they will be able to simulate anything in the universe on a computer, it simply being a matter of putting in the right information in the first place. Once this has been achieved we will then be in the position of having a virtual universe running on a computer. That such a possibility exists has prompted some people to question if our universe is nothing more than a virtual universe running on some super-being's computer. This is however, nothing more than comparing the universe to the latest technology that we happen to have, and has no bearing on reality.

Pythagoras lived in the sixth century B.C. His school, of philosophers, the Pythagorans, were convinced that the cosmos was based on numerical relationships. Pythagoras discovered the mathematical relationship between the lengths of strings that produced harmonic tones, the octave, for example, corresponded to the ratio 2:1. This Pythagorean connection between musical notes and the harmony of the cosmos was expressed by the assertion that the astronomical spheres gave forth music as they turned - the music of the spheres.

Kepler, in the scientific era, described God as a geometer, and in his analysis of the solar system was influenced by the numbers involved, and gave them a mystical significance.

Newton's view of the universe was one of a Designer working through strict mathematical laws. For Newton the universe was a vast and magnificent machine constructed by God. He envisioned it as a kind of perfect clockwork mechanism that had been wound up by God and was now relentlessly ticking away following a precisely determined path.

Today we compare it to a computer programme, tomorrow it will be something else. It doesn't matter, it's just our way of trying to comprehend the incomprehensible.

What is so amazing about mathematics is that it appears to be possible to represent the entire universe mathematically.

Einstein's theory of relativity and the theory of quantum mechanics, the two great theories of modern day physics, the most powerful, accurate and descriptive theories of the universe, were both developed mathematically. Why is the universe so mathematical? Is it that mathematics - at some deep level - is an essential ingredient of the very fabric of the universe?   Or is the universe nothing more than an expression of mathematics, the reality of mathematics, the ultimate truth?   Some say that God must be a mathematician, but instead it may be the other way around, it may be that mathematics -being the ultimate and eternal expression of truth and perfection - is God.

Acknowledgements:
John Barrow *"Pi in the Sky"*, Julian Brown *"What's in a Number?"*, Paul Davis *"The Mind of God"*, Roger Penrose *"The Emperor's New Mind"*, I. Peterson "*The Mathematical Tourist.*

CHAPTER 2

FAITH
and
PHILOSOPHY

13. Are all religions false?

"To believe in God or in a guiding force because someone tells you to is the height of stupidity. We are given senses to receive our information within. With our own eyes we see, and with our own skin we feel. With our intelligence, it is intended that we understand. But each person must puzzle it out for himself or herself."
Sophy Burnham

All religion is based on faith, and faith is just that, faith, it cannot be shown to be true or false, we believe what suits us, and there are a great many beliefs to choose from. We can even construct our own personal belief, and many do.

It is of course just as impossible to prove that all religions are false as it is to prove that religions are true. However, the following are my reasons for why I do not believe in any religion.

For the vast majority of people their religion is something that was forced upon them at birth and taught to them while at a very young and impressionable age. I do realise that some people find religion later in life as a mature adult, such as born again Christians for example, (note 'again') but they are very much in the minority in the overall world population. Some of us even reject religion as mature adults, even though we were raised in a religious belief. On average though, religion is, to the majority, a belief system that they were indoctrinated into, not one they chose when they were of an age to make their own decisions, because by that time they were already conditioned into the 'faith'. Religion is therefore very much to do with a person's society, culture and history and says little about a person's faith. Faith of course is real, lots of people have faith in one religion or another, it's the truth of religion that I call to question.

The question raised here is if any religion is true, or is the entire belief in religion and God (Gods), based on a falsehood or misconception, and that there is absolutely no basis in truth for any of it, other than as an entirely man made social construct.

Religions

Let us look at one religion in particular and see what we can learn. I have chosen Mormonism, for no other reason than that it is a recent religion and therefore its beginnings are easier to research because they are well documented. I assure you that I have no particular grudge against this religion, and to be fair, the Mormons that I have met on my doorstep have seemed to be pleasant enough decent people just doing their job. It can be argued that it does not matter how a religion was formed, but that the only thing that matters is the work that it carries out, and I am sure that the Mormons do carry out a great deal of good work.

Mormonism (The Church of Latter Day Saints) was founded by Joseph Smith, born 1805 in Sharon, Vermont. According to the official story of the LDS Church, in 1820 Smith was visited by God the Father and God the Son, after praying about which church he should join. He was told by Jesus, he should join none of them for they were "all wrong" and all the Christian church's doctrines "were an abomination" (Joseph Smith - History 19, Pearl of Great Price). This First Vision was followed by several visits from an angel named Moroni who, in 1827, directed Smith to retrieve a set of Golden Plates which had been buried in a hill near Smith's home in Palmyra, New York.

Between 1827-1830, Smith, with the help of friends, translated the Golden Plates into the LDS scripture, the Book of Mormon. Published in 1830, this was to become the first of many scriptures for the Mormon Church. By this time, Smith had also officially organised the LDS Church and was gaining a following.

The main point of interest here however, is how Joseph Smith started his new church. He claimed to have had a vision (he was the only one to have had this vision of course) that told him to start a new church with himself as the leader (how fortunate for Smith). He was told where to find the 'golden plates' that contained 'the scriptures' (they just happened to be buried, by an ancient civilisation, conveniently near his home in New York). He set out alone and 'found' them alone. (No witnesses). Once found, only he could translate them with the aid of a mystical device that only he was in possession of. (Very fortunate). Smith allowed no one else to see the plates and 'translated' them in secret from behind a screen. Okay, we've got the picture.

Smith was a known trickster and fraudster and was charged by the police on a number of occasions for various offences.

Some may say that the whole thing was an obvious fraud devised by Smith to set himself up as the leader of a new church. Indeed, to believe otherwise would take a monumental leap of faith that would include leaving behind all common sense and reason. BUT, Mormonism is a growing religion that has a massive following. Why? Because it was not 'Smith's' idea, it was revealed to Smith in a vision, or so he claimed. This is crucial. If he had attempted to start this new church as his idea, without such a 'devine seal of approval' the entire concept would never have got off the ground. People wouldn't have believed tricky Smith, but the words of Jesus, that's a different matter entirely. It could be argued however, that it really doesn't matter how the Mormon Church was founded as many would say it does a lot of good anyway.

All religions have the same basic ingredients, with local variations. Laws are introduced in order to control the population. These laws come from a higher authority that cannot be challenged because they come from a source beyond the reach, or question, of man. It keeps the population in check - follow the 'law' or be punished, obey and be rewarded, though not in this life of course which we admit is pretty horrible, but after you die and go to paradise. So go back to your grubby hovel, scratch a living from the dirt, bury your dead and don't bother us again with your stupid questions.

It can be no coincidence that the countries that have the poorest populations also have strongest and harshest religions. Take for example how the terrorists that flew the planes into the World Trade Centre and the Pentagon willingly sacrificed their lives in the sure and certain knowledge that by doing so they would enter paradise.

Without this belief I doubt if that many people could be persuaded to commit suicide for a political cause, but by making it a religious cause it completely changed everything, they believed they would not simply die, but instead enter paradise.

Death is another reason for the strength in the beliefs of religion. The vast majority of people are naturally afraid of death, it is the normal fear of the unknown. By its very nature we can never gain knowledge of what may befall us after death, it is our worst nightmare come true. We all know for certain that we will die but do not know when, how or why. Religion is there to comfort us on our trip into the great unknown, all will be well, in fact it will be far better than anything you could possibly imagine. Now that's a really good sales pitch, work on people's fears then sell them the solution. Its like showing a potential customer some really dreadful photographs of fire damaged houses then selling them a dozen smoke alarms.

My comments on religion however, are not to be confused with the way in which the church works within society. They do a huge amount of good work, some of them, and if that was all they did they would be a great organisation.

As for the holy bible, its just a book like any other, it contains pages of printed text bound together. It does not possess any mystical power of course, any more then a daily newspaper does. The content, however, is interesting in what it claims. It was once held as containing the absolute truth, but even this is being challenged. We no longer believe that the world was created in a few days. Modern church leaders are also saying that Mary, mother of Jesus, was not really a virgin, and so on. Back in the past people did genuinely believe the bible was 'gospel' and beyond question, but times, fortunately, have changed. The bible is being 'updated' so that it will again be acceptable to today's society. But is it really meaningful, can it be held up as 'proof' that God exists? No, of course not, we all know that, I may as well hold up a copy of 'UFO News" as 'proof' of the existence of flying saucers.

In all discussions on Christianity, I know that at some point someone will start to quote at me from the bible to back-up their argument. I really wish they wouldn't bother, its meaningless to do so, because it requires that the bible be accepted as hard evidence, and as we have seen it quite simply isn't so. The bible can be interpreted any way you want, so why bother? And precisely how accurate are the stories in it anyway?

It is very difficult, if not impossible, to be completely objective about religion because we have all been raised to believe in one religion or another. How do you examine the subject without some of your own bias creeping in? I, for instance, was raised to believe in the Christian faith, simply because that happens to be the majority religion in the country, that by pure chance, I was born in. I didn't choose Christianity, I was less than a year old at the time I became a signed-up member of the church. I certainly do not believe in it now. I do not believe that God is a kindly looking, bearded, white gentleman sitting on his throne in heaven surrounded by angels, and yes, that is the picture we were taught to believe as young, impressionable children at Sunday school, along with a whole load of other stuff just as daft. As an adult, all this was revised and we were told it was just symbolic, but as an adult I was also told that only true church going Christians will enter the Kingdom of Heaven. So to all you believers of other religions, sorry, you won't be going to heaven, our local vicar told me. Clearly, Christianity is just as ridiculous in its own way as Mormonism.

"God is love"
I saw a typical poster outside a church recently depicting an idyllic country scene, complete with pretty waterfall, a little mist, pretty flowers, lambs and song birds, bearing the slogan "God is love". Why is it that those who believe that God is responsible for the way the world is, only show the good things? Would it not be just as correct to show say a picture of thousands of corpses as a result of a severe drought - and include the slogan 'God is love'? The

Keith Mayes

church only promotes the one (good) side of God's supposed work. It is therefore, nothing more than a brainwashing propaganda exercise. The church must think us daft, and in order to believe their cozy, rosy picture of the world with good old father God taking care of things, we would have to be.

If 'God' can be praised for any good event that happens in the world, then this is claiming that God is responsible for the way the world is. That being the case God must also be responsible for the bad things as well, so why not attribute them to Him? If the bad things are not due to God, who says so? God? To blame mankind for these 'bad' events can not be right. Is it our fault that we have droughts, storms, floods, volcanic eruptions, heat waves, ice ages, earthquakes, tsunamis, hurricanes, tornadoes, disease, etc. etc.? I find the way that the church "cherry pick" 'God's work' to be extremely annoying. If its good, its the work of God, if its bad its just one of those things. Why not just accept the fact that events are the result of chance, cause and effect, and random factors? If you do believe God is responsible for the way the world is then as well as praising Him for the good things you would also have to blame Him for the headline events listed on the following poster.

GOD IS LOVE

Earthquake: Thousands die

Drought: Thousands die

Crop failure: Thousands die

Heat Wave: Thousands die

Flood: Thousands die

Big Freeze: Thousands die

142

However, the Church does have an answer to explain all the bad things that happen to us, it is the work of the devil, of Satan. How convenient! What puzzles me is that if God is all powerful why does he allow the devil to do all these terrible things when he has the power to prevent it? To my way of thinking because God stands back and allows the devil to do his evil work then God is as bad as the devil.

All religion is nothing more than an attempt to control a population by enforcing 'laws' that come from a higher authority that is beyond challenge, It is used as an extremely powerful tool to keep people 'in line'. It is particularly useful in circumstances where the population are under duress, be it poverty, starvation, tyranny. whatever, because then the powers that rule will assure you that you must accept your lot, lead a good life, don't rock the boat and cause any trouble, and you will be rewarded in the next life. Oh, good! The more you apply cold logic to it, the more senseless and totally meaningless the whole charade becomes, except as a means of control.

If religion is nothing more than a man made device for controlling the people, why do people believe in it? Because they want to! Religion gives meaning to people's lives. See "Is there a reason for our existence?" It gives a sense of security and purpose. It helps people to accept the dreadful and terrible things that happen in this world, such as mass starvation, earthquakes, fire, flooding, illness, etc. because its 'all part of God's plan' and so on. It also helps them cope with the prospect of death and the loss of loved ones. This belief however, obviously does not prevent any of these events from happening, and believers and non-believers alike suffer the same deaths.

Imagine for a minute that you are God with all his unlimited powers.

1) You decide that the people on planet Earth need guidance for their own good. Do you:

a) Appear before one person in a vision, give them secret instructions and hope they pass it on correctly and don't take liberties.
b) Appear before the masses and tell them direct, thereby proving your existence beyond doubt and at the same time ensuring that they get the right message.

2) You wish to have your laws continued forever. Do you:

a) Tell them once and disappear for a couple of thousand years or more.
b) Make regular appearances to reinforce belief.

3) You claim to care for these people, so when you see things going terribly wrong. Do you:

a) Sit back and watch the fun.
b) Step in again with some timely advice.

4) You have unlimited powers, you can create the world any way you want. Do you:

a) Make the climate changeable and unpredictable resulting in the deaths of millions due to drought, floods, heat waves etc.
b) Create a stable climate so that the population can grow their crops and feed themselves.

For all questions answered
a) Really bad idea, award yourself nil points.
b) Well done! good choice, you get full marks.

So why then do we have all the above wrong answer situations in the world and none of the good answers? If you believe there is a God, how do you answer that? Why would He act in such an uncaring and nonsensical way? Please do not reply with that old standby phrase reserved for all tricky questions, 'God's work is beyond our understanding'. You can bet your life it is! Given unlimited powers any one of us could make a much better job of it. It could of course be argued that God does exist but does not interfere in the affairs of us mere mortals. I think that would come as a bitter disappointment for all those that pray to God. Take away this belief that there is a God, in whichever religious guise you happen to prefer Him, and our existence seems rather nasty and pointless doesn't it. See "Why does the universe exist?" The universe is reduced to a cold unfeeling place that we just happen to have evolved in, that we struggle to survive in, and that has nobody out there looking after us because we are totally alone. When we die its all over for us, there's nothing more, and our descendants continue the eternal struggle for survival. We are nothing more than animals that happened to get smart and are here by pure chance alone, and by the same token at some point in the future we may not be here at all.

By removing God and religion we would merely throw away our security blanket, but nothing would really change, apart from society, and that must surely change for the better without religious divides. Life would continue as before and end as before. It would create a temporary problem for all those who's theories call upon God to have created the universe, but that would soon change.

Our world is not perfect and never will be, so why not get used to it, because that's the way it is. Religion never has, nor ever will, solve anything. Only we can do that. As for religion, its just make-believe, a heart warming myth to pacify and control the masses who otherwise would not be quite so happy to accept their miserable lot, or their eventual demise. However, we would all be better off without it because no matter how unpalatable the truth may be, its better than living a lie.

What evidence is there for GOD?

If you were to ask me if I believed in God then in order to answer your question I would require that you first define what you meant by 'God', otherwise we could be talking at cross purposes. Once you had done that I could then answer if I believe in the same concept of God that you do. I think that's obvious.

In order to continue on a sensible footing, let's outline a 'sketch' of God that we can all relate to in general terms. This God created the universe, His son, Jesus, was born in Bethlehem, the mother of Jesus was Mary. God and Jesus both performed miracles. Jesus died on the cross. God is in heaven.

I won't add more to this definition because the more I define God the more chance there is that we will disagree on some minor point or other. The idea is to simply agree on a very broad definition in order that we will be of the same mind when we talk about 'God'.

So why do people believe in God? This is a tough one, very complex. The main reason may well be because we were raised to believe in God. Then there is the church of course, reinforcing this belief, and the Holy Bible. Also add peer pressure- conformity within our social group, and just wanting to believe in order to give life a purpose or meaning.

What evidence do we have? In the purely scientific sense, none of course, but what can we take as 'supporting evidence' in a more general sense?

God created the universe?

Not really acceptable - see "Where did the universe come from?" In a nutshell this rests on my argument that if it is acceptable to propose that God has always existed (and therefore required no creator) then it is equally acceptable to propose that the universe has always existed and required no creator, thus eliminating at a stroke the need for a creation event.

The Holy Bible?

A collection of writings of various sorts from many different sources over a period of hundreds of years. Nonetheless an impressive historical record. Not to be taken too literally of course, it was written in the language of the time for the people of the time, but interpreted with a modern approach to its meaning still yields much first class historical information. It is clear that a good many people believed Jesus to be the son of God and that he performed miracles. That Jesus actually existed seems highly probable. As to his being the son of God, that's another matter entirely. Many, many people over the centuries have made that claim, or claimed to be prophets of God, it doesn't mean they were. Jesus claimed that he was the son of God, but it doesn't mean he was, or that he even believed it himself. That thousands of his followers believed he was the son of God doesn't mean he was. People of that era believed in prophets and prophecies, they were desperately waiting for events to happen, just as prophesied. It is known that Jesus deliberately staged events in order to be seen as fulfilling prophecies. They saw 'signs' in everything which encouraged their expectations, a classic example of a 'self fulfilling prophecy'. The bible is very persuasive, but not proof, of the existence of God, but a very strong case for arguing for the existence of a man called Jesus.

Miracles?

Let's put these into two categories, those in the bible and modern day miracles. Biblical miracles took place over 2000 years ago, you either believe they happened or you don't. We can't go back and 'test' them. Today's miracles? What miracles? These come in many forms, ranging from 'impossible' recovery from terminal illnesses to 'narrow escapes'. The term 'miracle' is applied too loosely and too freely. A miracle, as defined in Collins New English Dictionary is: "a super-natural happening". Recovering from illnesses that had obviously been incorrectly diagnosed as terminal is not that uncommon, it is not 'super-natural', it is misdiagnosis and/or lack of knowledge of the bodies ability to heal itself. The fact that the recovery took place after a visit to Lourdes does not alter the fact that the 'dying' person would have recovered anyway if they had instead stayed at home and watched the Simpsons.

We sometimes see on the news heart warming stories of survival against the odds, where a loved one is found safe and well after having thought to have died in some terrible accident or other. We often hear that well worn tedious quote "It was a miracle, God saved me", or words to that effect. Do these people who make these remarks have any idea of what that statement really means and what the implications are? I very much doubt it, for if they did I think they would remain silent! What they are saying is that God saw their plight and quite correctly decided to intervene, spare them all that nasty pain and suffering stuff and prevent their untimely demise. How very decent of Him. So little Jimmy was saved by God and thus was the only survivor of an aeroplane crash in which every other passenger and crew member died, all 326 of them. Even if little Jimmy was a detestable nasty little brat I am sure that the vast majority of us would be very glad to see the young lad survive the crash. Good old God we cry! So does the fact that little Jimmy was saved mean that it was in fact an Act of God, a miracle? I would think of it not as a miracle but class it as pure luck. However, some would claim it as a miracle. This type of incident is often claimed to be a miracle because the odds of surviving that particular accident are considered to be very remote. Using this logic we can now say that a miracle is all down to probabilities, in other words a mere statistical probability event. If we ignore the fact that highly improbable acts are classed as miracles (why isn't winning the lottery classed as a miracle when the chances of winning it in the UK are roughly 49,000,000 to one?) we are still left with a major problem regarding miracles.

The problem is *why did God save little Jimmy yet allow everyone else on the aeroplane to die?* As He intervened anyway in order to save little Jimmy why not save them all? And what about the thousands of innocent children around the world who die a slow and painful death by starvation every single day? Why save only little Jimmy? The same could be said for the nearly 3000 killed in the World Trade Centre attack, why did God save little Jimmy but stand back and allow them to die? Was He busy polishing His nails at the time? It is not the lone 'miraculous' survivor that is of interest here, the point is why allow all the others to die?

If there was a God I am sure He would not act in such a cold hearted and irrational manner. If on the other hand there is a God, I would like to have a word with Him and I most certainly could never bring myself to worship such a callous and cruel being. It may be of course that God has a policy of total non-intervention, that being the case there is no such thing as miracles.

What has happened to all those 'spectacular' miracles of 2000 years ago? Why did they stop? The seas do not part anymore to allow refugees to cross, the starving masses are not fed with a few fish. Why not?

I can not bring myself to believe in miracles. Either then or now. Also, think about it from a physical point of view. If God created the universe and all the laws of nature, do you think it possible that He can change those laws once they are in place? In just one location, for a brief period of time, and then change them back again? And all this having no lasting effect? The universe carries on just as before as if nothing had happened? Ok. Ok. I know what you're going to say, He can do anything! Personally I doubt that very much. If He can, why not feed the starving masses? He was happy to intervene in the past, according to the bible. Miracles of this sort, 'real' miracles, have not taken place for over 2000 years, if we are to believe they ever took place at all. Odd that.

I realise that I am directly at odds with none less than the Pope, the Vatican and the entire Roman Catholic church with my views on miracles being total nonsense. The Roman Catholic church has a rather bizarre and antiquated system for granting sainthood upon those who began life as mere mortals. There are a number of hoops to jump through before becoming a saint, the first of which is that you must be dead. It's no good being very ill, or even terminally ill, or very nearly dead, you must be actually dead, very dead, departed, deceased, devoid of life. This first and essential requirement is beyond argument and cannot be challenged under any circumstances whatsoever, you simply have to die before you can even be considered for sainthood. Having achieved that stage, and that's the easy bit (still interested?), you then have to be beatified, and in order to do this it must be proven that you have interceded with God on behalf of those who pray to you. In other words, people have prayed to you to perform a miracle - 'please make my puppy better' - and you, being a very decent sort of corpse - had a word with God who agreed that the miracle could go ahead. The puppy makes a miraculous recovery and you get the credit. You're on your way!

Now here comes the next stage. Can you believe that a bunch of old men in the Vatican, collectively known as The Vatican's Congregation for Saints' Causes, huddle together and pour over the 'evidence' as to whether or not a miracle actually took place? I wonder how on Earth they decide? If somebody recovers from a 'terminal' illness why is that considered to be a miracle when it is only an unexpected recovery after having been, obviously, wrongly diagnosed? It does happen from time to time, doctors are not perfect, anyone can make a mistake.

As an example of the sort of thing The Vatican's Congregation for Saints' Causes argue over, we can take the case of Antonio Gaudi the architect who died under the wheels of a tram 77 years ago. The Vatican's Congregation for Saints' Causes is considering a petition to beatify the devout architect. A pile of documents gathered by the Archbishop of Barcelona, offered as proof of Gaudi's ability to intercede with God on behalf of those who pray to him, is in the hands of the Vatican for it to pass judgement on his putative saintliness.

Among them are claims that the *'twisting, soaring towers and eccentric, colourful, ceramic adornments that liven the Sagrada Familia's exterior* "have the power to convert unbelievers! His backers are confident that, sooner or later, Gaudi will become the blessed Gaudi, and be placed on the first of what are several rungs of the ladder to sainthood.

Don't you find that absolutely ludicrous? Gaudi is most likely going to become a saint because "adornments that liven the Sagrada Familia's exterior have the power to convert unbelievers." Amazing! I really am finding it very difficult to take these people seriously, they must be joking, surely? If there was a God, I am sure he would look at what the Roman Catholic church are doing in His name and be horrified. All this religious nonsense, all those crazy rules made up by the Church, are man made and have nothing to do with God. We are in the 21st century now, and its about time that the churches got their acts together and started to move with the times. They could make a good start by dropping all this saint and miracle nonsense.

Perhaps there was a God once, perhaps there is now, perhaps there isn't. Maybe there never was one. Maybe it's all just wishful thinking. Perhaps if God ever did exist He doesn't exist now. Perhaps He died.

Has God Died? As to whether or not God existed in the past is open to debate. But is He here now? There is no indication that He is. No messages booming down from the sky, no miracles, no sign whatsoever. Why not? If He expects us to believe in Him, surely it is not asking too much that He gives us a sign every once in a while? Is it? To expect us to take it on trust for over two thousand years is asking a bit much. Perhaps He hasn't died, perhaps He can't, so where is He then? I suppose He could be busy elsewhere, making another universe or something. Doesn't seem very God-like to abandon us in this way though does it, we could use a little help just now.

Heaven? I am going to raise a final question regarding God. If you believe in God then you presumably believe in heaven. Here is the question based on the belief that only good people go to heaven, as per the bible, or those that truly repent. Would someone born with a brain tumour that eventually caused them to become insane and commit murder be allowed into heaven? If not, why not? They didn't choose to have a brain tumour. At what stage of development do we 'qualify' for judgement? At birth before we have actually done anything? At one year of age or 22 weeks in the womb? Define 'bad'. It's all a bit arbitrary isn't it.

Heaven is something that I just can not accept as a sensible concept. Hope I'm right!

I have received a number of emails via my web site informing me that I am incorrect when I say that only good people go to heaven. Sinners can in fact go to heaven, apparently, providing they repent. No, it's not good enough just to say you are sorry, you have to really mean it from the bottom of your heart, God can tell you know! So on this basis we have a new scenario. Imagine that Hitler, responsible for the extermination of six million Jews, finally realises the error of his ways, and in the week before his death feels sorry for killing all those innocent men, women and children - not to mention all the unspeakable experiments - and repents. All is now forgiven in the eyes of God - hell, we all make mistakes - and Hitler is allowed through the Pearly Gates. How nice. Meanwhile, a vicar, a man who has lead a blameless life, dedicating himself to helping others, makes a mistake, he sins. For once in his entire life he commits a sin, just once. Let's say he succumbs to the advances of an attractive female parishioner who just happens to be married. Our man does not repent because he says he is only a man, made of flesh and blood like any other, and has his weaknesses like any other, and is the way God made him. So he does not repent and when he dies is not allowed into heaven, where Hitler is currently residing and having a wonderful time. This is supposed to make sense? This is just? It's just more religious nonsense.

Are we 'hard wired' to believe in God?

It seems strange that many people across the entire globe, from all sorts of backgrounds and different upbringings, believe in a God of some sort or another. A great many people believe in God. Why? Perhaps science

Science, The Universe And God
The Search for Truth

can come to the answer, even in this area which is all about faith, not facts.

According to Dr. Michael Persinger it could all be down to Temporal Lobe Epilepsy and is researched in an area known as Neurotheology. Dr. Persinger is a neuropsychologist at Canada's Laurentian University in Sudbury, Ontario. His theory is that the sensation described as "having a religious experience" is merely a side effect of our bicameral (having two chambers) brain's feverish activities. Simplified considerably, the idea goes like so: When the right hemisphere of the brain, the seat of emotion, is stimulated in the cerebral region presumed to control notions of self, and then the left hemisphere, the seat of language, is called upon to make sense of this non-existent entity, the mind generates a 'sensed presence'. It is this 'sensed presence' that we refer to as either God, a ghost, an invisible presence, someone watching us, the presence of a departed loved one, or even leading to alien abduction stories.

Dr. Persinger has carried out over 900 tests on volunteers stimulating their brain to try and reproduce in the laboratory the feeling of 'sensed presence'. This is achieved by placing a helmet over their heads that contains electromagnetic field-emitting solenoids on the sides, aimed directly at the temples. Commands are typed into a computer outside the test chamber, and selected electromagnetic fields begin affecting the brain's temporal lobes. The lobes are bathed with precise wavelength patterns that are supposed to affect the mind in a stunning way, artificially inducing the sensation that the subject is seeing God. It may seem odd to reduce God to a few ornery synapses, but modern neuroscience isn't shy about defining our most sacred notions - love, joy, altruism, pity - as nothing more than static from our impressively large cerebrums. Persinger goes one step further. His work practically constitutes a Grand Unified Theory of the Otherworldly: He believes cerebral fritzing is responsible for almost anything one might describe as paranormal - aliens, heavenly apparitions, past-life sensations, near-death experiences, awareness of the soul, you name it.

Dr. Persinger suggests that naturally occurring magnetic fields in certain areas may be responsible for frequent reports of ghosts, or religious visions, or UFO's, whatever. From his research he has discovered that some people are far more receptive to these changes in magnetic fields than others, and those that are most receptive tend to be the most religious. By way of an experiment Persinger tested Prof. Richard Dawkins of the University of Oxford, a well known and respected scientists, author of many books and well known atheist. Dawkins was keen to participate and was actually looking forward to experiencing a 'religious experience', but was eventually disappointed as he experienced no sensations out of the ordinary. Dawkins, it transpires, is very insensitive to magnetic fields - as shown in the test results - which does tend to support Persinger's theory.

At the other extreme to Dawkins is the case of the teenage girl who experienced such strong feelings of 'presence' in her bedroom she was unable to sleep at night. Tests revealed that her electric alarm clock, that sat close to her head, emitted a strong magnetic field, and once it was removed from her bedroom she did not have any further problems.

Could it be true that our sense of religion, our belief in God, our sense of 'otherworldly presence' is nothing more than the effect of magnetic fields interfering with our temporal lobes? Persinger seems to think so, and he could be right.

What do I *really* think?

I'm not really sure, sometimes I believe, sometimes I don't. Mostly I don't. Almost always I don't, but just sometimes I have this little nagging doubt, and that's because I was taught as a young child to believe in God.

So I ask myself this question: "Had I been raised in a secluded society where nobody believed in God or had ever heard of religion, would I, by my own experiences and philosophy, have come to believe in the existence of God?" The answer is, I believe, no. I would have no reason to. The next question I ask myself is: "Having been raised in total ignorance of any religion or God, when I die, would I go to heaven?" The answer to that, I think, is yes. If there is a God, why should He prevent me from going to heaven, providing I had led a decent life of course, just because no one had told me about it? Not my fault is it? If God refused to allow me into heaven that presumably means I would go to hell (I don't know of any in-between place) and I would consider that to be grossly unfair. And God is just! So I would go to heaven on appeal. He lets me in, just as He does all those people who attend church on a regular basis, pray, worship, donate money to the church, etc etc. Bet they wouldn't think it fair! Is there a court of appeal for admission to heaven?

So there you have it. You either believe or you don't. Logic has nothing to do with it.

It seems to me that it really does not matter one way or the other what we happen to believe, the truth is the truth regardless of belief. There are so many different religions to choose from, and if any of them are right it can only be one of them. All the others are wrong. The mathematical odds dictate that you have probably chosen the wrong one! Does that make you a bad person?

14. Is there life after death?

"Life is pleasant. Death is peaceful. It's the transition that's troublesome."
Isaac Asimov

One of the great reasons for believing in one religion or another is that all of them assure us of an after-life, a paradise, a wonderful place where we will rejoin our loved ones, in one form or another, providing of course we have the faith and live by the code! But is it true? How can we ever know?

A great many people do believe in life after death, but is that because they have arrived at that conclusion by using their common sense, or is it because that is what they have been taught?

First of all let's be clear what is meant by death. In the normal stages of death the heart stops beating and respiration ceases. Without a supply of oxygen the bodies various organs stop functioning and start to die. The brain, also starved of oxygen begins to shut down. Death occurs when the brain ceases to register activity. All straight forward stuff.

So the difference between being alive and being dead is activity in the brain. We are our brain. What is the very essence of us, what makes me me and you you, is our brain. So what is it?

The human brain is a collection of cells that for the typical adult weighs about 3 lbs. It contains 100 billion neurons. The neurons form connections to each other called synapses. These synapses produce chemicals called neurotransmitters. There are 1,000,000,000,000,000 synapses in the brain. A neuron is activated and the synapses connects to another neuron via a neurotransmitter through its synapses. Every thought you have, every movement you take, every memory you possess, is nothing more than a particular sequence of synapses. This is what we are, a collection of neurons sparking off a chain reaction to other neurons. When the brain is no longer activating neurons, we cease to exist.

The rest of our body serves no other purpose than to keep the brain alive and to reproduce.

So if that's all we are, then to answer the question 'is there life after death?' the answer would have to be no. The real question then is, is that all we are?

Do we have a soul?

Those of a religious nature would argue that we are far more than that, they would argue that we have a soul. A soul is, for me, hard to imagine. How do you describe it? It is not something that we can detect, it is a spiritual thing without any physical substance, and it is this that supposedly lives on after our mortal bodies have died. So where is the soul while we are alive? I assume it must reside in the brain as that is where we 'reside'. I can only assume again that when the brain dies the soul is released. That being the case the soul is able to survive without the need of a body, it is obviously independent of the body. So why does it need a body in the first place? Perhaps it needs a body to develop and is unable to leave it until the body dies.

According to popular belief only human beings possess a soul, animals apparently do not. So what is the difference between humans and animals? As far as I can tell it's just a question of intelligence, as measured by us. After all it's not that long ago that we were animals, we weren't created human, we just evolved along different lines than the others. That being the case there is nothing special about us that should warrant us having a soul and animals not, after all, some people are born with severe brain malfunctions that on an intelligence scale would rate them less than some animals. So intelligence can not be the deciding factor, unless you believe that you need to score above a certain IQ level in order to be granted a soul. If that were the case, and we tinkered around with genetics and produce a chimp with the IQ of a schoolboy, would that give it a soul? Would the chimp go to heaven?

So what exactly could it be that determines whether or not a life form merits a soul or not, as the case may be? I think we can rule out intelligence. What else sets us apart from the animals? Some would say our appreciation of the arts, poetry, music, etc. I don't agree. I think that just comes with intelligence. What else is there? Absolutely nothing at all! We think we have a soul, we think we are 'superior' to the animals, we think there is some undefinable quality about us that sets as apart from the animals. I don't think so. We are just animals that got smart.

I do appreciate that there are people who believe in the biblical version of the creation of Adam and Eve by God in the Garden of Eden. That is their choice, but in this day and age it has to be said that that idea goes against all the evidence so painstakingly accumulated by palaeontologists over hundreds of years.

I have been contacted via my web site to be informed that I am wrong to say that animals do not have souls and that all life forms have a soul. If that is the case then I wonder where we draw the line for what organisms qualify for a soul? Dogs and cats? Rats and mice? How about birds, flies, worms, maggots? Then we have bacterium, or how about a flu virus? Where do we stop? Plants? Does plant life count? Again, as with the case of deciding if a thing is good or evil, it's all a bit arbitrary.

What I find hard to accept is the belief that somehow the human race is special, more than just an animal. We are animals, we evolved out of the same primeval swamp as the rest of the animal kingdom.

Getting smart is not in my view a passport to gaining a soul.

If you wanted to convince me that we do have a soul, you would need to explain what it is that qualifies us for a soul and not animals. And if animals have a soul, then where is the line drawn?

Ultimately though, being a matter of faith, and faith alone, you either believe it or you don't. There is not a single shred of proof that we possess a soul. Some though would argue that ghosts are lost souls.

Ghosts

Another reason for believing in life after death is the belief in ghosts, the belief in a spirit world. Many people claim to have seen ghosts, some to have photographed them or caught them on video tape or on audio tape. Ghosts are sometimes described as being 'lost souls', or the souls of people that are unable to 'leave' due to some unfinished business, or some such thing. So are ghosts real? Or is it all just a trick of the light and an over active imagination?

It has to be said that the human brain has a wonderful capacity for seeing patterns where none exist. The 'ink-blot test' is a good example of the ability of the brain to see patterns and shapes in a random mixing of colours. It should therefore come as no surprise that some people, in a highly charged or tense atmosphere, who are expecting to see ghosts, will see ghosts. The thing that concerns me is that when people claim to have seen or photographed a ghost, is that it is in direct contradiction of what a ghost is supposed to be. A ghost is claimed to be a spirit, a soul, and not a physical thing, so how can it be seen? We cannot see, or photograph an emotion. You cannot see or photograph love, hate or happiness. We can only see material, physical things, not spiritual things. So how can it be possible to see a ghost? As people claim to see ghosts walk through walls, and other solid physical structures, they cannot possibly have any material existence. so therefore we cannot possibly see them! If, on the other hand, a ghost has a physical reality that enables us to see it (how then does it walk through walls?) then it cannot have anything to do with a soul, the two - ghost and soul - must be unrelated. A soul cannot possibly have a physical existence, that would also be a contradiction in terms. It also strikes me as very odd that ghosts are usually reported as wearing clothes! Are we to assume from this that clothes have a soul?

The only conclusion I can draw from this is that ghosts and souls are not related, they are two different things. Therefore ghosts cannot be used to support any argument that there is life after death, or that we have a soul. What ghosts may be I have no idea, but strongly suspect that they exist only in the imagination of those that claim to have seen them. The world of spiritualist, seances and the like, belong to that segment of the population that believe because they want to. For all the talk, all the hype, all the 'photographs', all the wonderful stories, there is not one shred of evidence that ghosts exist.

Near death experiences
One of the reasons sometimes raised for believing in life after death is the controversial topic of 'near death experiences', NDE's as they have become known. This is where a patient 'dies' on the operating table but is subsequently revived. They often report that they felt surrounded by peace and love, that they were calm and relaxed. They report going through a tunnel towards the light, a feeling of being drawn towards the light. They also report 'out of body' experiences where they rise above their own body and can see and hear what is happening around them. I do not intend to enter into a long and involved study of NDE's, there are many books on the subject and an even greater number of web sites for those that are interested. All I will say is that a great many people are desperate to believe, for whatever reason, that there is life after death. Perhaps they are afraid to die?

I have not had an NDE, nor do I particularly want to have one, but I did once think I was about to die. I had a nasty and lasting chest pain, I had woken that morning with it, and still had it mid-afternoon so went to my doctor. I was aged about 50 at the time. My doctor gave me an ECG and after looking at the print out for a few seconds immediately telephoned for an ambulance. Within a short period of time I was laying in the back of an ambulance on my way to hospital with the siren going, wired up to an ECG, with a paramedic looking at the monitor and anxiously looking at me. I could hear the machine beeping out my heart beats, and it was missing an awful lot of beats. For the first time in my life I thought I was about to die. I had often wondered what it would feel like, knowing you were facing your imminent death, now I know. I had always expected that I would feel fear, the fear of the unknown, the fear of facing death, but I didn't. It wasn't at all like that. I felt angry! I felt annoyed that I was going to die and that I wasn't ready yet. I hadn't finished living! It was a feeling of helplessness and of annoyance, fear just didn't come into it. So if you have been worried at all about facing your death, don't be. Most people report a feeling of relief, of happiness. Some, like me, a feeling of annoyance, but scary it isn't!

History

Throughout history many cultures have expressed their belief in an after-life. Perhaps the most well known example is that of the ancient Egyptians and their pyramids.

Why do so many cultures believe in an after life? I think the answer to that can be found in the natural cycle of nature, Every night the Sun is seen to set, yet every morning it is seen to re-appear, to be re-born. Winter is a time when many plants to appear to die, but come the spring they bloom again. Every year the seasons repeat themselves in and endless cycle.

The ancient Egyptians were first class astronomers, and used the positions of the stars as an annual calendar to predict when the Nile would flood. They noticed how all the stars appeared to revolve around one central part of the sky, the North Celestial Pole. This area of sky was always visible, whereas the other stars would periodically disappear below the horizon. They considered this part of the sky to be immortal, never changing, a very special place. They believed that by gaining entry into this eternal place, they would live for ever in the after-life. This is why the pyramids were precisely aligned with the cardinal north/south co-ordinates as opposed to the magnetic north. They believed that the pyramids would transport the Pharaohs to this immortal place in the sky and would thus ensure that they would live forever.

The ancient Egyptians therefore believed in an after-life because of the natural cycles seen in nature, and because of their knowledge of astronomy, and their belief in Gods. History is not therefore a good guide when it comes to supporting a belief in life after death, they knew even less about things in the past than we know today!

A reason for our existence?

Another reason for belief in the after-life is that it gives our lives a meaning, a purpose, that otherwise would be lacking. Many would argue that if we simply lived and died, and that was all there was to it, then we would have served no purpose. They see our 'purpose' in life as developing our soul, preparing the way for our next life in the here-after.

I have a number of issues with this way of thinking. I cannot understand why people think that we must have a purpose. Did the flea that I just trod on exist for a special reason? Does the planet Pluto exist to serve a particular purpose? Was I created to fulfil a particular purpose? Were You? Or is everything that exists simply the result of the initial conditions of the Big Bang creation of the universe? I do not believe that we are here for a 'purpose', or to prepare ourselves for the after-life. I believe that we are just here, period.

As this is rather involved and complex subject, I have given it its own section, See "Is there a reason for our existence?"

To sum up. Is there life after death? Nobody knows. What evidence is there to support the belief? None? What reasons are there for believing? Many, take your pick.

Does it make any difference what we think? I don't know, do you?

15. Is there a reason for our existence?

"I have too much respect for the idea of God to make it responsible for such an absurd world."
Georges Duhamel (1884 - 1966)

Many people believe that there is a reason for our existence and usually, though not always, it is based in religion. See "Are all religions false?". This may be a very profound belief that the only reason that we are here is because we are somehow required to be in order that we may serve some purpose in a grand plan of some description or other. This belief may take the form that this existence of ours is merely a brief step, or even a 'test' before we move on to greater things in another plane, another life, or indeed in Heaven. It is a very hard belief to ignore, even for those who would not normally describe themselves as being of a religious nature, but which can I think be explained by our own egotistical nature.

We all have of course self awareness, a feeling of 'self', of being aware of ourselves and our surroundings, and the only clear view that we can have of the world is, naturally, our own. We see the world as a sequence of events that happen to us and around us. We of course, from our own perspective, are very much central to all these events. It is not therefore surprising that we each consider our own existence to be important, regardless as to the actual reality of the situation. Logic would dictate that if one of us was to cease to exist the world would carry on pretty much as it always has, apart from being missed by friends and family of course. But logic can be ignored when it suits us and the vast majority of us consider ourselves to be important to 'the great scheme of things' to some degree or other. Our egos are not keen to accept that in reality we may not be of any consequence to the vast universe in which we exist, or indeed to humanity as a whole. Our importance to family and friends is, naturally, accepted.

Following on from the belief of our own self importance it is, I suggest, only a natural step to assume that we are here for a reason, to serve some purpose, even though we may not know what it may be. I am not talking here of family commitment or of our duty to others, but a grander purpose in the 'great scheme of things'. But is it true? Is there really a need for us to be here? Do we have a purpose within this vast universe?

If we apply logic to the question, then if we are here for a reason, we have to conclude that our existence must be necessary, otherwise we wouldn't be here. That being the case our existence alone should satisfy the 'need' to be here because we do not ourselves know what that reason is or what it is that we are required to do. It would follow that if our existence comes to an end, then we have 'done our duty', whatever it was. In other words, whatever we do, or don't do, however long or brief our lives, it was all part of the 'plan'. It seems to me then that no matter what we do, how we live our lives, it was meant to be, that was the 'plan', providing of course that you believe in a 'plan' in the first place.

On the other hand if we believe that we are here by chance alone, then again it doesn't matter to the 'plan' how we live our lives, because we don't believe that there is a 'plan' and possibly do not believe in God either for the same reasons. It will of course matter to others how we conduct ourselves. Great 'plan' or not it would appear that we are free to act as we see fit. This is assuming of course that we do in fact have free will in our choice of actions, and there is no proof that we do. See "Free will and existence".

There is however another way to look at the idea of a plan, a greater goal for the human race, that would give us a purpose, a reason for being, other than as described above which is based on each individual having their own reason for existence. We could take the view that the plan, or reason, applies to the human race as a whole, and that the actions, or even the existence, of individuals is unimportant to the overall 'direction' of the plan.

By way of example let's consider the route taken by billions of electrons following a path of least resistance. They reach a 'gate' that allows 30% to go left and 70% to go right. The path splits at the 'gate' and 30% go left and 70% go right as the 'gate' intended. The 'gate' works on probability and does not, indeed can not, choose which individual electrons go left and right, it has no bearing on the result anyway, it's only the percentage that go left or right that matter. When observing the electrons approaching the 'gate' it is impossible to predict which of the electrons will go left or right. The laws of Quantum Mechanics work on probabilities, but given a high enough number of electrons the outcome of such a 'gate' can be predicted with remarkable accuracy. So the selected percentage of electrons go where they were designed to go and our computer works.

We can imagine ourselves following a 'route', performing a function just like the electrons flowing through a circuit board. The electrons of course are unaware that they are part of a highly complex system and are performing remarkable feats of number crunching so that we can read a web page for example, they are merely existing and flowing along a circuit from which they have no escape, or even a notion of escape. The existence of individual electrons is unimportant, the route they 'choose' to take is unimportant, and in fact each individual electron does indeed 'choose' its own route. It is only the overall outcome of the statistical probabilities that is important, that is what makes the computer work, and the electrons of course have no notion of a circuit board, let alone a computer. Could we be like those electrons? Could it be that individually we are of no significance, that our individual actions count for nothing, that it matters not if we live or die, but collectively, as the human race, we do have a reason for existence, a collective goal for mankind that is beyond our comprehension?

Another way of looking at the problem is to take a more holistic view of the universe and everything that it contains. Everything is made of the same original elementary forces that coalesced from the big bang singularity, whether it be a photon, a star, a lump of rock or a human being. It is only the combination, the mix of things, that makes things different from one another. When we peer millions of light years into the depths of space, we still see the same things, further back in time of course, as we find in our own galaxy and here on Earth. We, the human race, are as much a part of the universe as a spiral galaxy or black hole, and made of the same stuff. Is it then fanciful to suggest that we, as sentient beings examining the universe, represent the universe examining itself? Because in a manner of speaking it is. We are not 'different' to the universe, we are very much an integral part of the universe as a whole, we may be only a very small part, but we are a part of it, we do not exist in isolation. So next time you look up at a dark sky and see those distant twinkling stars just remember that it was in those fiery furnaces that the first steps were taking in building up the atoms that eventually led to you. We are made of star dust, we are part of the universe. Perhaps we should be asking instead why does the universe exist?

Throughout history there have been great people that have made sacrifices for the benefit of others, they had a cause and made it the reason for their existence, even sacrificing their lives for the sake of their cause. Perhaps they died content knowing that they had made a significant contribution to the advancement of mankind. Some would have believed that this was the only reason they were born, to do this great deed. They were sure they knew what was required of them by whichever God they happened to worship. They may be right, but what would have actually been achieved? A better life for those remaining, yes, for which all are grateful, but in 'the great scheme of things' merely a more comfortable existence for those that were going to exist anyway, (according to the 'plan') not a real change in the 'plan' itself, life goes on regardless, but towards what we do not know. Even the belief that we are 'heading somewhere' is entirely without substance, unless you believe in a 'plan' for reasons of faith.

I do not believe there is any reason for our existence, and of this I feel certain. Why? For the following reasons:

1) Life has evolved on this planet because the conditions just happen to be suitable. It did not require a 'design plan'. Pure chance. We just happened to have evolved into what we are today by chance, natural selection, mutation, survival of the fittest etc. etc. and just plain luck that we were not destroyed by some calamity along the way. I do not consider that events were designed in such a way that our development was a necessary outcome, I think it just happened and equally may not have. As our existence appears to me to be the result of a chance sequence of events the idea that we are are here for a reason would be, as Spock would put it, illogical.

The initial conditions in the Big Bang all those billions of years ago started a chain of events that resulted in our existence today, but that is not a plan, that is just a natural consequence of those initial conditions. Even if you believe that God created and designed the Big Bang it is not necessary to believe that He is controlling our lives. Why should He? He doesn't need to. See 2) below. If you believe that God created the universe for our benefit, that we are the reason for the creation of the universe, do we, on planet Earth, really require the existence of 100 billion galaxies just for our own survival? We have many times more to the nth degree then we could possibly need in our own galaxy alone, the Milky Way, and the chance of us reaching even the nearest stars in that seems incredibly remote. Why then such a staggeringly ridiculous excess? God could have just made our Milky Way Galaxy.

2) We could all be destroyed at any time by an asteroid or comet, just like the dinosaurs were believed to have been 65 million years ago, see "Asteroid impacts:can we survive? ". Next time it may be the entire planet that is destroyed. What would have been the point of our existence then? You could I suppose argue that by then we will have served our purpose, whatever it may have been, and somehow 'moved on' as it were. What purpose? This would indicate that we are being 'overseen' by some higher authority. God

perhaps, Who will decide when the time is right and wipe us out with, say, a comet.

If there is a God that is so all powerful and is in some way controlling and tinkering with our lives, then why make us go through all the terrible things that happen on this planet every single day? With so much power why not just make us as He intends us to eventually become? Why make us imperfect or incomplete in some way, put us on the planet to live and die and so 'improve' ourselves, when He could just make us perfect in the first place? Whatever it is that we are here to learn or experience as a 'necessary' process before progressing to something else as part of the 'plan', God could simply just 'give' to us in the 'next stage' and bypass this stage altogether. Save a lot of time and trouble.

3) To believe in a 'plan' that would give us a reason for our existence, does, I believe, require us to believe in the existence of God, for only God could be the creator of such a 'plan'. If you do not believe in God then you presumably do not believe in a plan. I say 'presumably' because perhaps there is a way at looking at the development of mankind towards some final goal that does not require that God play any part in it. But I consider that would be taking an altruistic view of life, believing that we should act for the benefit of mankind, striving to make things better, which is a wonderful view of life to take, a great goal to aim for, but is that a reason for our existence? It is only improving our existence, not a reason for it.

To sum up. A 'plan' that would give us a reason for being here? No, I do not think so. I do not see any logical point to having a 'plan'. A 'plan' of this magnitude could only be created by God, and He does not need a plan, He can make it happen, anyway, anytime He desires. I do not however rule out the possibility of God, only that He is not responsible for the way we run our lives, that's up to us and chance. He may have created the universe but He is surely not controlling our individual lives. I would hate to think that He was, because if so, could He not make it so that nobody gets sick or starves to death? I'm sure He could of course,

He can do anything, so the fact that these things happen suggest to me that if God exists He does not play any part in our day to day lives. I would also suggest that if He does exist He plays no part in the workings of the universe at all. So without some form of 'plan' - and I am sure for all the above reasons that there isn't one - then we are not here for any reason at all. We just exist.

As I said in "Where did the universe come from?" I do not believe we need to introduce the concept of God into the equation to explain how the universe may have come into existence, and I furthermore do not believe we are here for any purpose of God's. If you however believe the opposite to be true, then all well and good, but why do you think God needs a plan?

If you think that there are areas that we are somehow not fit to ask questions about, then why do we posses the intelligence to do so? It could have been 'blocked off'. We have a brain and I believe we should use it to question everything, even the reason for having an intelligent brain, we must have it for a reason, don't you think?

Is there a reason for the existence of the universe?

Suffice it to say that if the universe had a creator, then the creator alone would know the reason, but this should not prevent us from attempting to discover what that might be, if indeed there is a creator and a reason. I tend to take the view that if there was a creator then the creator must have had a reason, even if only out of idle curiosity to see what would happen! The creator may have a Grand Plan for the universe, but why? He (or She) does not need a plan, they can just make it happen.

Perhaps though I am being over generous with the creator's powers. For example, It is a simple matter for computer programmers to devise a program, with all the initial rules built in, but have no idea how it will 'turn out'. To illustrate, there is a computer game called "Life". In this game squares are marked out on a grid similar to a large chess board. Different 'beings' live in the squares. They are programed to move to other squares depending on the

169

condition of their surrounding squares and on there own condition. Some, when landing in an occupied square marry and have children, some fight and kill, some die if they are surrounded, some evolve into something else, etc. It's a bit like real life. The start of the game involves placing beings in some of the squares then seeing what happens. It is actually impossible to predict the outcome of the game, it is not known until it is finished, after a set number of moves. The ebb and flow of 'civilisations', births and deaths, forms intriguing patterns across the screen, but can not be predicted. The computer, when given the initial conditions is unable to predict the outcome without actually running the program! So even though the rules are known and the initial conditions are chosen, the outcome is unknown, it only reveals itself as the game unfolds.

Could the universe be like that? Why not! Simply because the universe may have had a creator does not require that the creator knew how the universe would develop. The eventual 'outcome' of the universe would be the same, creator or not, once the laws of physics are set (the rules) and the initial conditions of the Big Bang chosen ('pieces' and 'values' of the pieces), then the game is started in the Big Bang and things take their natural course. (See "Where did the universe come from?" regarding the Big Bang.) The universe heads for its destiny, but what that may be could be completely unknown, even to its creator. In this scenario, there would not be a reason for the existence of the universe, except as a kind of experiment, it would not exist to serve a purpose, it would just exist.

If we take the view that the universe did not have a creator, then the question of there being a reason does not arise, unless you consider that the universe contains its own reason for having an existence.

I do not believe the universe has a reason for existing, I believe it just exists. There could be an infinite number of universes existing, some of which may have developed life, others which may not have. We should not assume that 'ours' is the only universe, we should not attach so much importance to ourselves, it is rather like the earlier belief that the universe revolved around the Earth. There is however still an 'in built' belief that we are somehow 'special', and I feel that this largely stems from a belief in God and that He put us here for a reason. Isn't it amazing the number of scientific theories where God creeps into the argument! It used to be that for everything that could not be explained it was accredited to God, or The Gods, from crop failure, disease, abnormal weather conditions, and a host of other events. As knowledge has progressed, and answers found, The need to blame, or thank God, has steadily declined as we look to science for the answers. See "Are all religions false?".

We do not often have to call upon God in order to explain natural events these days, but in some areas that we are still seeking answers, some people still do. I am not saying that they are wrong, but I am saying that perhaps they are being too hasty looking to God as a 'final solution' because history has shown that given time, we often do find the answers. Perhaps one day we will have all the answers. Perhaps when that day comes we shall all be as Gods.

16. Will we become immortal?

"I would never die for my beliefs because I might be wrong."
Bertrand Russell

If at the time of the end of the universe we are still flesh and blood, the answer is clear, we will die with the universe, whether that be in the Big Crunch or by perpetual expansion and heat death of the universe. But can we imagine a scenario in which we are no longer flesh and blood? In order to explore this possibility let's first examine how we became what we are today and then attempt to extrapolate from that what we may become.

At the moment of creation of the universe - the Big Bang - the universe consisted of a cosmic fireball of radiation at a temperature of 1,000 billion degrees Kelvin. Matter could not exist at this temperature, the universe was far too hot for anything but radiation. As the universe expanded and cooled the first atoms of helium and hydrogen were formed. Once formed, the processes began that led to the formation of galaxies, stars, planets and life, according to the physical laws of the universe. See "Is the Big Bang theory correct?". Life on Earth started with very simple single cells that gradually evolved into the vast multitude and complexity of life that we have on Earth today.

It could be argued that life started with the single cell, but it must be remembered that the single cell has its beginnings firmly rooted in that initial sea of radiation in the cosmic fireball, as does everything else in the universe. We can trace our own beginnings all the way back to that radiation. That's how we started out in the universe, we were the universe. What we have evolved into, intelligent life, allows us to examine the universe and ourselves as two separate entities, even though we come from the same beginning. We see ourselves as existing in the universe, not as part of it, but we are. See "Is there a reason for our existence?".   This is where we are today, what of the future?

The future is at best always a guess, and I would guess that physically we won't change that much. In the past it has always been Darwin's survival of the fittest and climate changes that have shaped our evolution, but these factors just don't have the same effect on us today. Fitness is no longer a requirement for survival of the hunter-killer, we can simply drive to the supermarket and fill a trolley with food, which is just as well for most of us. We are well adapted to the environment, and when it becomes unsuitable we simply modify our environment (air conditioning, central heating, etc). Physically then, not much is likely to change, so in this respect we are all doomed.

Our only hope therefore is with technology. Could we possibly become clever enough to manipulate the universe in time to prevent its death? Seems a tall order, but not to be completely dismissed.   The major problem with this however, is that the only energy available to us must come from within the universe, we can't pop outside and bring some in. The energy required to control the expansion of the universe must be at least equal to the total energy possessed by the universe.

What then will become of us? Will we ever become immortal, or will the human race simply cease to be? If we are unable to control the universe, perhaps we can control ourselves?   Could we use technology to give us immortality?

Our physical bodies do not last long, and even with transplants and prosthetics, we cannot expect to live forever, everything wears out eventually. I think that our future development depends upon our intelligence, it is the one thing that separates us from all other life on this planet, and this may indicate the path our future development could take.

Predicting future outcomes however, is nothing more than guesswork, and is often wrong. Nobody for example foresaw the development of the silicon chip, and that changed everything. Take a look at old black and white science fiction films, they show space ships being operated by dials and levers. In today's films the dials and levers have been replaced by computer screens and keyboards (Star Trek) or by voice activated self-aware intelligent computers (HAL in "2001:A Space Odyssey"), You can imagine in say fifty years time they will look back at '2001' and laugh because we had not foreseen...whatever. Trying to predict next year's developments in computer technology is virtually impossible, trying to predict events thousands of years into future is clearly impossible.

However, computers may offer an interesting possibility. We are already at the stage where we can communicate with computers by voice. My iMac for example, understands and performs my verbal instructions, most of the time, and on occasion will 'verbally' inform me of the stupidity of my instructions. Military jets are being developed that will respond to the pilot's thoughts and will launch missiles and perform manoeuvres via thought commands through a special helmet that is wired into the cockpit. The next step will most likely do away with the helmet. I think the time is not that far off when we will all be able to directly interact with the computer by thought alone via a transmitter/receiver micro-chip in our heads, no wires required. I understand that this work is already in the development stage. No monitor or keyboard will be required, interaction being directly from our heads to the computer and vice versa. As for the next stage in computer development, as I said earlier, that's impossible to know, but we could make a guess or two......

So far in this section, as in all other pages, I have kept to known and near technology, and stuck to the facts, but what comes next can only be pure guesswork. I have tried to imagine what could happen, but it's no more than food for thought, so think on......this is just for FUN.

Computer hardware will become ever smaller and will eventually no longer require silicon chips and hard drives.

Instead they will consist of nothing more than a few carefully arranged atoms that will operate at the quantum level, (See "What is Quantum Theory?") as will the 'chip' in our heads. (Quantum computers are already in development, but at this early stage still require a lot of hardware) We will no longer see computers and will barely be aware of them, but will interact with them constantly in our day to day lives through our quantum chips. Everybody fitted with the quantum chip will have access to almost instant information and to each other.   Advances in technology will accelerate at an unbelievable pace. We will reach a stage where it will be impossible to differentiate between us and computers, we will be virtually be one and the same thing and no longer even consider their existence.   Eventually computers will not even have an independent existence, we will have completely absorbed them into our quantum chip as they become so small. What we will have become is no longer contained within our organic brain - that now merely controls our bodily functions - but we will have become our quantum chip, because all our thoughts, memories and thinking processes will be contained within it. All those with the 'chip' will eventually become to think as one. The next advance is almost inevitable, the transfer of our quantum chip data, our thoughts, memories - ourselves - into a machine. We each become a robot, a machine made of metal, plastics and electronics.

We now exist inside a machine, no more flesh and blood, we have no further need of it. The body would have served its ultimate purpose, which was the creation of an intelligent brain.   The brain had only one purpose, the existence of its intelligence independent of the body with all its inherent weaknesses. Life has only one goal - to survive - and the ultimate form of survival must be independence from the vulnerability of flesh and blood. We become a form of computer, not a computer, but something far greater than a computer could ever become. See "Will computers become self-aware?".   You could perhaps call it an intelligent self-aware super computer, only it isn't, because it's us. We never succeeded in making an intelligent self-

Keith Mayes

aware computer, we became it, and in the process went far beyond that.

From within our artificial containers we collectively learn to adjust to our new form of existence, and without the limitations imposed upon us by our physical bodies are at last finally free to think without fear or restraint. The result is the final stage of our evolution, the ability to make the last transfer, out from the constraints of those few remaining atoms that bind us, and into the vastness of the universe. We 'download' into the quantum virtual particles that are the very fabric of the universe. This becomes our final home.

We will have come full circle and become the universe again, only now the universe is intelligent and self-aware. As the universe now possesses control, it will not come to an end, it's far too intelligent to allow that to happen.

So we become immortal, but along the way we leave behind the very essence of what is was that made us human. To answer the original question, will the human race become immortal, the answer is no, it will not. It is the universe that will survive, and that was the plan all along. The 'plan' did not exist at the beginning, at the Big Bang, because the universe could not have a plan without intelligence. Having gained intelligence it gained the 'plan', and with it the means to have caused its own creation.

The universe was created because it would eventually develop intelligence that would bring about its own creation. John Wheeler "In search of the edge of time" and "Schrodinger's Kittens" takes the Copenhagen Interpretation of quantum theory at face value and applies it to the entire universe. He suggests that the wave function of the entire universe - right back to the Big Bang - only collapsed into reality once the universe was observed, and this observation brought both it and ourselves into existence. See "What is Quantum Theory?

Fanciful? yes, of course it is, but as I said, it is food for thought isn't it? And a bit of harmless fun. But is it possible? Your guess is as good a mine, but why not? No

176

doubt though the truth, as always, will be far stranger. Anyway, who's to say it couldn't happen?

I expect someone will.

Regarding the progress made in the world of computers, and how small they are becoming, the following article may be of some interest, even though it is already a little dated.

PRESS RELEASE
Date Released: Monday, November 26, 2001
Weizmann Institute

A Trillion Computers In A Drop Of Water- Scientists build a nanoscale computing machine using biological molecules.

A group of scientists headed by Prof. Ehud Shapiro at the Weizmann Institute of Science has used biological molecules to create a tiny computer - a programmable two-state, two-symbol finite automaton - in a test tube. Reported today in Nature, this biological nanocomputer is so small that a trillion (1,000,000,000,000) such computers co-exist and compute in parallel, in a drop the size of 1/10 of a milliliter of watery solution held at room temperature. Collectively, the computers perform a billion operations per second with greater than 99.8% accuracy per operation while requiring less than a billionth of a Watt of power. This study may lead to future computers that can operate within the human body, interacting with its biochemical environment to yield far-reaching biological and pharmaceutical applications.

The computer's input, output, and 'software' are made up of DNA molecules. For 'hardware,' the computer uses two naturally occurring enzymes that manipulate DNA. When mixed together in a solution, the software and hardware molecules operate in harmony on the input molecule to create the output molecule, forming a simple mathematical computing machine, known as a finite automaton. This nanocomputer can be programmed to perform several simple tasks by choosing different software molecules to be mixed in solution. For instance, it can detect whether, in an input molecule encoding a list made of 0's and 1's, all the 0's precede all the 1's.

'The living cell contains incredible molecular machines that manipulate information-encoding molecules such as DNA and RNA in ways that are fundamentally very similar to computation,' says Prof. Shapiro of the Institute's Computer Science and Applied Mathematics Department and the Biological Chemistry Department. 'Since we don't know how to effectively modify these machines or create new ones just yet, the trick is to find naturally existing machines that, when combined, can be steered to actually compute.'

Two more articles, this time published in Focus Magazine No.127 June 2003.

Chip for the brain
University of southern California, los Angeles.

US researchers have created a chip for the brain, which could replace the memory centre in patients affected by epilepsy or Alzheimer's disease. The silicon chip contains a mathematical model of the hippocampus, the brain's memory processor, and could one day be used to download memories into humans. Tests on the chip will be soon carried out on rats. The research is controversial because it targets areas of the brain that affect mood and consciousness.

Circuits from proteins
Technische Universitat Munchen, Germany.

The basic building blocks of life could be used to manufacture tomorrow's miniature machines. Experts in nanotechnology from the US and germany believes that complex arrangements of protein fibres could be moulded to create the next generation of microscopic electronic circuits. The researchers took a prion protein, which is several thousand times thinner than a human hair, and coated it with a layer of gold.

Seems like my 'silly' idea may have some merit after all!

17. Free will, perception and reality

"In the distant future I see open fields for far more important researches. Psychology will be based on a new foundation, that of the necessary acquirement of mental power and capacity by gradation. Light will be thrown on the origin of man and his history."
Charles Darwin *"The Origin of Species"*

Do we have free will?

I touched upon the subject of free will in "Is there a reason for our existence?" and in "What is Time?" and decided the subject merited further examination.

It seems to us that we have free will. We could decide to do this or that as we see fit. For example, I could have decided not to publish this book but chose to do so. But how much choice did I actually have? Was it my decision or was it already 'fixed' and it just seems to me that I chose to do it?

I have no way of knowing if I have a free choice of actions. There is no test that I can perform to prove I have free will, no matter what course of action I 'choose' to take I can not prove it was my choice. Just because it feels that we have free will does not mean that we do, we can't take it for granted based on a 'feeling'.

So how can we prove it? We can't. Like it or not we simply cannot prove that we have a choice of actions. If we go back to the instant of the Big Bang, those initial conditions led to things being the way they are today. The formation of galaxies, stars and planets, the emergence of life on Earth, were all determined by those initial conditions. Once the Big Bang started events happened that led to us being here. Could things have happened differently? That's debatable, but if the Big Bang were repeated, with the identical initial conditions, you would expect that, by and large, things would develop pretty much the same, perhaps exactly the same. Would this lead to such a close match that you and I would also be repeated? If we assume that the formation of the galaxies, stars and planets must be the same because the initial conditions are the same, then you would also expect the formation of life to be the same. At this point however, the rules change.

The formation of the universe would be the same because it would be following the same laws of physics as laid down in those initial conditions. Life however, once it has developed intelligence does not follow those same laws. Yes, we live and die by those laws, but we don't follow them, we change them. We have changed the surface of this planet for example. We change things, we cause things to happen that have no direct bearing on the initial conditions. These changes could not have been determined at the start of the universe, they were not 'built in'. We have irrigated desserts, reclaimed land that was under the sea, destroyed countless acres of forests, farmed the land, etc. etc. Planet Earth is no longer in the condition it would have been if left to nature. We have changed it by an act of will.

I suppose it could be argued that the changes that we have made are a part of those initial conditions, that the very way in which life has evolved has 'forced' us along this route, that it could have been 'calculated' from the initial conditions that human life would have evolved on this planet and changed things the way we have. Although it may seem ridiculous at first glance, it could be so.

If we take the situation where we are now, it would seem ridiculous that all the atoms in my body could come together in such a complex and organised way as to make me, from those original atoms in a far away star somewhere. But it didn't just happen like that, it was a gradual step by step process over millions of years, each simple step leading to a slightly more complex stage that gradually formed simple cell life and so on all the way up to me. Each step of the way was only a very small step forward, building on what had happened before. Maybe you and I are but another simple, but inevitable, step along the pre-programmed route.

It could also be that whether or not we have free will is not dependent on the initial conditions, but whether or not we are all following 'a plan' as I mentioned in "Is there a reason for our existence?"

So how much freedom does the individual have? We simply don't know. I cannot conceive of any test that would prove free will one way or the other. We just assume we have a free choice of actions because it feels as though we do.

My personal view on the matter is that we do have a free choice of actions. I do not think it possible the we are all following a pre-programmed course as laid down at the instant of the Big Bang. Could it have been predicted from those initial conditions that tonight I would go to Burger King with my wife, have a flame grilled Whopper then go on to the cinema to watch 'Lord of the Rings'? I think not. However, we can't be certain.

It just makes me a little uneasy that I can't prove it. You would think it a simple thing to prove, but how can you?

Free will is something that we tend to take for granted, just as we do the world round us. The sky is blue and the grass is green. But is it? Is the way the universe appears to me the same as the way it appears to you? This depends on how we each individually perceive the universe.

What is perception?
Imagine that you become shipwrecked and eventually reach a small desert island. You find the island is occupied by one other person who himself was shipwrecked some time previously. He is from Japan, you are from New York. You do not speak the same language, but by using signs and gestures you are eventually able to converse. To pass the time, your new friend whose name is Lee, describes his home and family. He points to a brown tree trunk and says that is the colour of his wife's eyes. He points to your red shorts and tells you that is the colour of his car, and so on. After a time you have built up a picture in your mind of Lee's

home and family. Eventually you are both rescued and you promise Lee that you will visit him in Japan.

The day arrives and you visit Lee as promised. You are not at all surprised to find that Lee's wife does indeed have brown eyes, Lee's car is red, and so on. By the power of language Lee was able to communicate all these colours to you. By using the tree, your shorts, and all the other items available, you had handy 'colour charts' that you could both relate to. You have a pleasant stay and then return home, confident in the ability of one person being able to accurately describe colour to another.

What you never knew was that Lee was colour blind, he could only see in monochrome. Every colour to him was either black, white or a shade of grey. Lee was not even aware that he was colour blind, he thought that the shade of grey he saw your shorts to be was exactly the same shade of grey that you saw them to be. So in reality, you had no idea of the colours that Lee was trying to describe to you that he had in his mind. What does this tell us?

It tells us about perception. Perception is simply how we, as individuals, translate the information our senses relay to our brain about the external world. To continue with the colour theme, for example. Visible light, as we know, is simply the range of electromagnetic radiation which human eyes are sensitive to, typically in the range of about 380 - 750 nanometres. At the limits of this range are ultraviolet radiation, (at shorter wavelengths), and infrared radiation (at longer wavelengths), neither of which are visible to our naked eye. A colour then is nothing more than a particular wavelength of radiation within that visible spectrum, blue for example being around 420 nanometres.

The reality of a shade of blue for example, is that it is radiating at the wavelength of 420nm. How I perceive it however is known only to me. The text colour of this page is to me black. I have absolutely no idea of how you perceive it, none whatsoever, even though you would also describe it as black. You may be colour blind and see it as a shade of, what I would describe as, blue. On the other hand you may have 'perfect' eyesight, but your brain may perceive it to be a shade darker, or a shade lighter, or as something completely different, compared to how I perceive it to be. There is no reason at all why our brains should all perceive the same things the same way. Our experiences of life differ, our first experience of a particular object, how it is 'translated' and 'catalogued' by our brain, is unique to us.

There is absolutely no possible way for us to communicate to one another how we each perceive the world. My world, the one that my brain has processed from the input from my five senses, is in all probability very different to your world. We can both look at a red rose for instance and agree that it is a red rose, but that's just putting a handy label on an object, it does not convey anything at all about our perception of it.

Reality is the actual physical universe that we exist in. The universe is the way it is regardless of how we perceive it to be. A truth is a truth, it doesn't require that you believe it is true to be so. Perception, however, is how we as individuals, translate that reality into information that we can use in order to relate to the external world. In other words, we all have our own personal version of reality that we carry around in our heads, and are totally unable to relate to any other person's version of reality, or them to ours. But it doesn't change reality.

Sad isn't it, that I will never know how my wife perceives the dozen red roses I give her on our wedding anniversary, to her they may be what I would describe as yellow. I wonder what they smell like to her? Does she hear the Atlantic rollers crashing onto the beach the same way that I do? Or Beethoven's Fifth? When I hold her hand, how does it feel to her? I will never know. We can never know what anyone else in the world is feeling, seeing, hearing, smelling or tasting.

We are, each of us, in our perception of the world, both unique and totally alone.

So what exactly is reality?

Reality, as defined by Chambers Concise Dictionary, is "The state or fact of being real: that which is real and not imaginary: truth: verity: the fixed permanent nature of real property (*law*)." How can we translate that definition from the pages of a dictionary out into the real world? What really is 'reality'.

One way is to compare what we believe to be reality with what we know that isn't. We can compare real life with fiction, with a film, and for this example we have an excellent film that will suit our purpose very well - The Matrix.

The Matrix, as I am sure you know, is a SF film based on the idea that the world we live in is no more than an illusion, and that in the real world - one that is hidden from us - we are existing in life support tanks and being used as a power source for our masters, machines with artificial intelligence. All very amusing and great special effects.

But could there be just a hint of truth to it?

Can we be sure that what we consider to be reality, really is reality? Or is it just a fleeting illusion, like the black dots in the grid below? The more we try to pin it down, the more elusive it becomes.

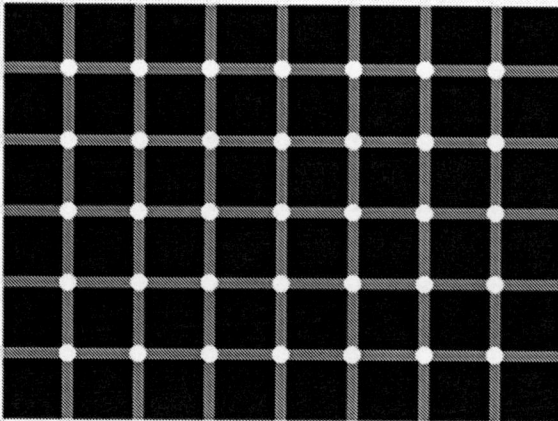

What test can we apply to determine if our reality really is reality? One way is to compare it with what we know to be an illusion, our dreams.

How do we know our dreams are an illusion? Because we compare them to what we call reality!

When dreaming we are not generally aware that we are dreaming, although on occasion we are. But when we awaken we always realise that we were dreaming, we realise that it wasn't real. We realise that nothing in our dream could actually hurt us, but does this mean it wasn't real? It seemed very real while we were dreaming it. They were real experiences taking place within our brain, the same brain that processes all our experiences, our only contact with the world around us. Dreams are so real to our brain - to us - that they can cause our heart rate to double, our blood pressure to rise, make us break out in a sweat, physically lash out against 'attackers', get out of breath and even make us scream - every physical reaction you would expect in real life. If our brain is experiencing it, and our bodies responding to it, doesn't that, for us, make it real? However, it comes back to what I said before, our dreams can't hurt us, only reality can. When our dream attacker hits us we do not have any injuries when we awaken. Why? Our bodies physically respond to our dreams, why not to what is inflicted upon it in our dreams? The answer is of course that as we were not 'really' hit we did not suffer any 'real' injuries, we only imagined that we did. From this we are able to deduce it was just a dream and not reality. If it were reality we would bear the injuries.

How else can we tell when we were dreaming? Very often by reflecting upon the events of the dream we can see how events were not logical, that impossible things happened. Thus when we awake and recall flying unaided through the air, leaping tall buildings in a single bound etc, we know that it was a dream and not a real event that happened yesterday. We can use common sense and logic to determine the obvious, to differentiate between dreams and reality. Dreams do not have the restraints of the physical laws of nature, follow the principle of cause and effect, and are not required to make any sense.

When dreaming we tend to believe it is reality, but when awake we know it was just a dream, we can distinguish it from reality. But dreams are a kind of reality, we cannot just dismiss them as though they do not exist, they do exist, dreaming is a real phenomenon, even though the dreams themselves are only an illusion. So perhaps dreaming can be considered a type of reality, if not actual reality.

Starting with dreams as a type of reality, our next step obviously is to call our 'real' reality, the reality we experience when awake, 'actual reality', not being aware of any stage in between being asleep and being awake. (The thought occurred to me of being in a coma, but as that is a form of damage to the brain I decided to ignore it.) We are now in the happy situation of knowing that our 'actual reality,' is the genuine article, that this really is reality, we know it is not a dream. Great. But......

But suppose this reality is just an illusion that only appears to be reality. After all, we think our dreams are real until we wake up. Could there be another - higher - level of consciousness that is 'the' reality? What test can we apply to prove or disprove this idea? The answer is that we cannot test this reality, we wouldn't know it wasn't the real thing until we 'woke up', as it were, and reached the real reality and were thus able to make the comparison.

Are there any grounds for believing that a higher level of consciousness exists, that there is a higher level of reality? Can reality have levels? If dreams can be considered a level of reality, then the answer must be yes, we can have levels of reality. Big 'if' there though, accepting dreams as a level of reality.

What else do we have that may support the idea of other realities? How about other dimensions? Theorists have long suggested that other dimensions must exist in our universe, See Can anything 'real' be infinite? Why not other levels of reality? The problem here is that other dimensions can (apparently) be proven mathematically, and perhaps experimentally eventually, but we cannot prove reality.

How about death? Could death be the gateway to the next level of consciousness, to the next level of reality? Could death be a kind of awakening? Is death the moment when we finally wake up to the truth? Wake up to the real world? This is becoming eerily close to how many religions describe death. Could this be the answer then?

Could what we are experiencing now be only one level of reality, of which there is another - higher level - the gateway to which is death? Could death be not the end of life, but the true beginning?

Many faiths would argue this is indeed the case. Perhaps they are right.

If then our lives are not the true reality, but only an illusion, a glimpse, a fraction, of the true reality, then just how important is this life of ours? See "Is there a reason for our existence?". We tend to cling onto life by every means at our disposal, because we fear death, because we fear the unknown. This is not surprising, not only is death the great unknown, it can also be preceded by pain and suffering, something that we are naturally programmed to avoid. It is perhaps the sensation of pain, or indeed any sensation, that convinces us that this is reality. But what if it isn't? What if this reality is nothing more than our starting place, where we are initially created, a necessary beginning which eventually leads us to our death, where we discover the true reality? Where we find an understanding of the timeless infinity that in this life is beyond our comprehension?

What does the bible have to say on the matter? I think that depends on how we interpret the bible. For example:

"And I saw a new heaven and new earth: for the first heaven and the first earth were passed away; and there was no more sea. "Revelation, 21.1

"And God shall wipe away all tears from their eyes; and there shall be no more death, neither sorrow, nor crying, neither shall there be any more pain; for the former things are passed away." Revelation, 21.4

Maybe we should look at religion in a different light? Admittedly, the bible quotations above are not meant to be taken as our individual passing from this life to the next, but in the more general context of judgement day. In other words perhaps, our 'real' death, not just our mortal demise.

If we view the bible as the literal truth, then we would have to believe that we were created by a superior being, as was the entire universe, and that when we die we shall merge as one with the universe, and all pain and suffering will end. Perhaps we are looking at existence and reality from the wrong perspective.

For example. Is it possible that when and if scientists actually manage to achieve their aim of creating a black hole singularity in a laboratory they will create a new universe? This is believed to be a possibility. Lets look at the bible again and see how it connects to cosmology.

"In the beginning God created the heaven and the earth" Genesis, 1.1

Is this then how God created the universe, by creating the singularity that created the Big Bang?

"I am Alpha and Omega, the beginning and the end, the first and the last." Rev.22.13 Is this a biblical description of the Big Bang singularity?

Perhaps by creating a black hole singularity *we* will create a universe that will spawn life that will regard us as God? Perhaps we are just existing in another being's black hole singularity created in a laboratory? God's laboratory.

Then there is the question of Virtual Reality, an artificial 'reality' created within a computer programme, consisting of nothing more than electrons pursuing paths along a specially designed and constructed circuit, as in The Matrix. How real is virtual reality? It's real, that's for sure, it does exist. But where does it exist? Only within the confines of a computer programme. Does this make it any less 'real' than our reality? Could our universe, designed and created by God, be termed a 'virtual reality' universe from God's perspective?

Would computers possessing true artificial intelligence be able to create virtual reality to the extent that the virtual creatures that 'live' within it consider themselves to be 'alive', just as the computer would consider itself to be 'alive'? (Define 'alive'). See "Will computers become self aware?" Could our universe be nothing more than a computer simulation and not a 'real' universe? Could we be no more than virtual beings who believe that we, and the universe, are real? Some very well respected scientists have suggested that the universe is far too well designed to be a lucky accident, that there has been some 'tinkering' with the laws of nature to make conditions so perfectly suited for the development of life. Could they be right? Could our universe be artificially constructed. Is this virtual reality?

This all comes down to how you define 'reality'. If God designed and created the universe, does not that make it artificially constructed, as opposed to one that developed naturally? Is this reality no more, or no less, real than 'artificial' or 'virtual' realities?

So just what exactly is reality? Which of the realities is the 'real' one? I think that just depends on the one you happen to exist in.

Is there more than our five senses are telling us?
Or are we missing something?

We are of course all familiar with the fact that we possess five senses. We can detect by touch, taste, sight, smell and sound. That's it, there isn't any other way for us to detect the environment around us. Evolution has decreed that we can get by with these five. Well, yes we can, but it would improve our survival chances if we had more.

Suppose for example we were able to detect when we were in a deadly radiation field, I think that would be quite handy, a sense that enabled us to detect radiation. I'm not talking about detecting it with the senses that we have, such as being able to see radiation for example, but a new sense organ that would require a new description of how it detected.

It's difficult, if not impossible, to imagine a new sense, we can only think only in the terms of the five we have, but try a new way of looking at things. Try to imagine that we had another organ for detection, and in the same way that we have eyes for sight, this new organ detects something that we were previously completely unaware of.

Try and imagine that we could detect the presence of a completely silent and odourless ambient temperature remote object in a dark room. With our usual five senses we couldn't, but with our new one we can. No, not by echo location such as a bat would use, that's just using sound and hearing, but a completely new method. We can detect it because its there. I'm not even talking about getting a picture of it in our heads by whatever means, this is different, a new awareness. We just know its there. This is a bit like attempting to explain colour to someone who was born sightless, but I'm sure you follow the idea.

I am going to digress here for a minute, but please bear with me, I am trying to make a point. Years ago I read this in a novel, so its not my idea. It was a pleasant summer's evening and I decided to do what I had just read. I went out into the countryside as the evening sky was darkening and laid down on my back on the grass looking up at the emerging stars. My feet were pointing towards the east. I picked out a bright star near the horizon that I could see through the branches of a nearby tree. I spread-eagled my arms and legs and gripped the ground with both hands as if I was holding onto the surface of a giant beachball and didn't want to fall off. I watched that star as it moved upwards through the branches of the tree. Now the important part. I imagined that the star was stationery and that it was the giant beach ball that I was holding onto that was rolling forward. The effect was amazing. It made me feel quite giddy, but all I was doing was experiencing things as they really are, the sky does not revolve around us, it is us that is revolving. Knowing it and being aware of it are two different things. It was the first time I had actually felt the sensation that the world was turning!

Why did I tell you this? Because it illustrates the way we accept things, the way we come to look at things without

191

so much as a second thought. Sometimes it is a good idea to look at things differently.

There is a popular misconception among some sighted people that when a person who was born blind feels an object they build up a picture of what it looks like. This isn't the case, they build up a plan of what it feels like, they are unable to mentally create a picture, they have never experienced pictures. If a blind person became sighted they would be totally unable to recognise a ball by looking at it for the first time, they would have to feel it first in order to recognise it, because they have no mental pictures with which to recognise it by sight. If you have a problem with that, imagine instead that you were born without the sense of smell. Would you be able to imagine the scent of a rose by touching it? That should make it clearer. A person born without one of the senses can have absolutely no conception whatsoever of what that missing sense is all about by simply using the other senses that they do possess. See "What is reality?"

So armed with a new sense, a new window on the universe, what would we 'see' that we don't see now? What things are we missing? What secrets would we uncover? It would be like giving sight to a blind man which opens up a whole new world. Even with artificial help we could never see into this 'invisible' world, we just see more than we could with our unaided senses. A machine can not present images to a blind man nor sound to one who is deaf.

Are there things going on that we are completely unaware of? I think there is. I don't know why, I just do. I have often wondered why it is that we have five senses, not six, not seven, but five. Don't you sometimes feel that you're wearing blinkers? Maybe it's just me, but are we missing something?

CHAPTER 3

CONSPIRACIES
and
CONFUSION

18. Did we land on the Moon?

"That's one small step for (a) man, one giant leap for mankind"
Neil Armstrong. Tranquillity Base 1969.

I can remember on the night of July 20th 1969 sitting in front of my television in the early hours of the morning watching a grainy black and white image of Neil Armstrong climbing carefully down the ladder of the Lunar Module and stepping out onto the surface of the Moon, the first man to have done so. It was a magical, awe inspiring moment that I shall never forget.   The space age had finally arrived. "That's one small step for (a) man, one giant leap for mankind". It certainly was.

There are now some people questioning if that event ever actually took place. They are basing this on what they perceive as some discrepancies in a handful of lunar photographs from among the 30,000 that were taken by the Apollo astronauts, and for various other assorted reasons. They believe that the Moon landings were mocked up in a film studio and that the Apollo crews never left Earth orbit.

The most surprising thing to me about the entire ridiculous claim that it was a hoax, is that some people actually manage to believe that it was a hoax! This is just about the most preposterous and ludicrous conspiracy theory ever invented. If it really was a hoax do you not think that the Russians at least would have been able to expose it? After all, the *only* reason for going to the Moon was to beat the Russians, yet they have never once even hinted at the possibility that it was a hoax, they know it was real! That is why they gave up their own attempt after they ran into a few problems with their booster. There was no point in them continuing once they realised the Americans had beaten them to it. Nobody in the world had more reason to want to prove it was a hoax then the Russians, but the Russians are not daft, they were closely monitoring the Americans every inch of the way, and were able to determine for a fact that the Americans did actually land on the Moon, much to their annoyance. If the Russians say the Americans landed on the Moon, then the Americans landed on the Moon, believe it!

Let's now move on to explain why all the reasons people believe it was a hoax are wrong. You do not need to be a rocket scientist to see the glaringly obvious errors in their arguments, common sense is all you need.

Let's start at the beginning with the first claim, that they never left low Earth orbit, before going onto the various photographs etc.

The Apollo crews never left low Earth orbit.

This is a ridiculous claim, that they never left low Earth orbit, because if that had being the case, how do you explain the rather obvious fact that not one person noticed that the Apollo craft were still continuously orbiting the Earth when they should have been orbiting the Moon? They would have been very easily visible to the naked eye, just as satellites are today every hour of the night. The Apollo craft were, by many magnitudes, the brightest and largest artificial objects orbiting the Earth, and would have been impossible to miss. Some people even actually saw the Trans-Lunar Injection burn for Apollo 8, from low Earth orbit to trans-lunar trajectory in the dark sky over Hawaii, and how could anyone fake all that?

I think it a bit odd, to say the least, that if they had stayed in low Earth orbit, that nobody in the world noticed a strange and very bright satellite that appeared at the moment of the launch of Apollo 8 and stayed in orbit until the moment of re-entry, when it mysteriously disappeared. An amazing coincidence that repeated with every Apollo moon launch, a further eight times!

The argument that they stayed in Earth orbit is based on the false belief that it is not possible to survive passage through the Van Allen radiation belts, but the danger is nowhere near as great as that. I gain the impression that hoax believers believe the Van Allen belts to contain the equivalent of alien death rays and nuclear explosions!

Do you honestly think that the entire world's radar systems, and visual astronomical observatories, let alone individuals, would have failed to notice them if they had stayed in Earth orbit? This would apply to all the manned missions that entered lunar orbit, Apollo's 8 to 17, a total of nine missions. And no one in the entire world noticed, not even once, that they didn't really leave Earth orbit? To suggest such a thing is ludicrous! No one would have been more over the Moon (sorry) to expose any American Moon hoax than the Russians, but they couldn't, they saw them leave Earth orbit!

This argument alone should provide sufficient evidence to show that the Apollo crews did actually leave low Earth orbit, but this will not be enough for the Moon hoax believers, they love a good conspiracy theory.

However, the only mystery about the Moon landings is why these people believe it was a hoax, but I really shouldn't be so surprised, some people even believe the Earth is flat. I can only assume though it is because they saw the TV program "Conspiracy Theory: Did We Land on the Moon?" It was shown In the States in February 2001 and later in the UK. The program points out 'errors' in the photographs, that there are no stars visible, that the flag waves in the 'breeze', and a host of other detail.

The manner in which the so called 'facts' were presented by the programme to support the hoax theory were so biased, unscientific and totally inaccurate, as to be laughable, and believing them makes as much sense as reading a comic to gain information on D.I.Y. brain surgery. This was just a run – of - the - mill sensationalist TV entertainment programme, the sort of programme that refuses to let the truth get in the way of a good story. Unfortunately though, some people will believe anything that involves a wacky conspiracy theory.

Photographic evidence.
Let's examine the claims behind five photographs that demonstrate how these errors arise.

1) NASA forgot to paint the stars in the sky.

This is a classic, my all time favourite. It is very popular with the hoax believers, but I can't understand why though, it's so easy to prove for yourself. I think it tells us something very important about the way they think. This is where we have photographs showing the Lunar surface, maybe an astronaut and the Lunar Module, and an area of black sky devoid of stars.

The hoax believers claim that NASA forgot to paint the stars in on the back drop in the studio where it was all filmed.

The real reason is that when contrasted with the brightness of the astronauts and the Lunar surface, the stars are just too dim to register on the photographic emulsion of the camera film. If the camera shutter were held open long enough for the stars to register, everything else would be over-exposed into a white featureless glare. You cannot have both visible on the one photograph, so the camera was set for the correct exposure for the lunar surface, not the stars. When standing on the Lunar surface the astronauts could not visually observe the stars in the dark sky, because of the surface glare, they could only see them when standing in shadow. By the same token, if we take a photograph outdoors at night from a brightly illuminated surface, our photograph also would not show any stars in the sky.

It is not enough that the Lunar sky is very dark, in order to see the stars you have to BE in a dark area yourself, and your camera. In 1967 Surveyor probes soft landed on the Moon and sent back amazing pictures of the surface. An image of the stars was required in order to learn the precise orientation of the probes. It took a three minute exposure before the stars became visible. The cameras used by the astronauts typically took images using an exposure time of 250th of a second. Not surprising is it that the stars did not register in the photos!

If it is so easy for hoax believers to spot this 'glaring error' - and let's face it, to forget to put the stars in would have been an incredibly stupid mistake to make - do you honestly believe that not one single person involved in the 'hoax' wouldn't have noticed it either? Or is it just that hoax believers are all just so much smarter than all of them?

Anyway, no need to take my word for it is there, I could be part of the conspiracy according to your way of thinking. (Wish I was, I would be getting paid for this.) Just pop outside one night and try to photograph the stars with a brightly illuminated person in the foreground. Try it, its easy enough to prove without the need of a massive conspiracy theory, just you and a camera is all that is required.

2) The Great Flag Waving in the Breeze hoax.

I just love this one, very nearly as much as the 'no stars' one.

This typically shows a photograph of the astronauts next to the stars and stripes, which appears to be flapping in the wind. The hoax believers claim that this is proof that it was filmed in a studio as there is no atmosphere on the Moon. They are very sharp at noticing small details like this.

The truth of the matter is that flag is held out in the unfurled position by an extendable rod running through the top of the flag, so that it can be viewed unfurled, and you can always see the unnatural rigidity this gives to the top of the flag in any of the photographs. The rod creates the effect of a breeze blowing the flag into that position. Without the supporting rod the flag would just hang limply down and would not reveal the stars and stripes. Flags are designed to be blown into position by the wind on Earth, so the support was added to replicate this, as there is no atmosphere on the Moon. In the most popular photograph the rod is not extended the full width of the flag and it looks like a breeze is causing a ripple in the flag.

It has also been claimed that some video clips show the flag waving in the breeze when it was planted. Not so. The movement of the flag is because when astronauts were planting the flagpole they rotated it back and forth to better penetrate the Lunar soil. Without an atmosphere it takes a while for this movement to damp down. There is not one video clip showing the flag moving when the astronauts are not holding it, a fact never mentioned by the hoax believers.

Do you really think that an errant breeze blowing through the set causing the flag to wave in what was supposed to be a total vacuum would not have been noticed? Such an obvious fact could not escape the notice of an entire film crew, besides which they would surely have called upon the services of experts to oversee operations to guard against this very sort of 'error'. They would simply have done another take.

3) The cross hairs have been added after and go behind some objects

I must admit to being rather fond of this one as well, as it is such a totally pointless 'hoax' that I fail to understand why anyone can believe it was actually done. The most foolish aspect of this claim is that if the cross hairs were added after, how can they possibly be overlaid on the photo and appear behind some of the objects in the photo? Hoax believers are defeating their own argument with this one!

The cross hairs on some photographs appear to go behind the objects in the photograph. The cross hairs are included as an aid for linking a series of pictures together to create a panoramic view. The cross hairs on the photographs were produced by a glass plate within the camera, between the lens and film. They result in a black cross on the film because they block the light from reaching the film directly below them. If, however, you are taking a photograph of a really bright white object, the over-exposed part of the film 'bleeds' into other parts of the film. This is particularly the case if the adjacent part of the film is black. This is exactly what is happening where the cross hair meets a bright, reflective part of the photograph. It occurs in a number of the Apollo photographs, but you only see it where the cross hairs seem to disappear behind a bright white part. You never see it happening anywhere else.

Why do you think NASA would want to add the cross hairs after? If they had somehow changed cameras and forgot to insert the etched glass plate that produces the cross hairs, they would have just ditched the photographs, not gone to the trouble of faking them in afterwards. It would be a hell of a lot easier just to do a re-take if the photographs were considered to be important enough.

4) The background has been changed and its supposed to be the same place

This one I feel is especially stupid. The hoax believers point out two photographs that NASA say are from the same area and argue that they can't be because the lander can still be seen but the background is different. They do agree that the camera has moved a little for the second shot, but not enough to explain the different background, therefore it must be a hoax because someone changed the background when they shouldn't have. The other argument is in reverse. This is where they say that NASA claim the photographs to be from two different missions, i.e. two different sites, but the backgrounds are the same. The problem with this second claim is twofold. One, they do not say where NASA claim the two sites to be different, and two, hoax believers are not above faking the photographs. No doubt this will cause howls of outrage from hoax believers, they couldn't possibly lie and fake photographs, only NASA do that!

So we will restrict ourselves to the first claim as it is the only one we can deal with by direct testing. The problem that hoax believers raise here is that they do not believe that the background can change significantly when the camera only moves a few feet, or so they assume anyway. So why don't they pop outside with a camera and test it for themselves. No, too obvious isn't it?

Below are two photographs that I took of my wife, with Loch Lomond in the background and some mountains on the horizon. I moved a distance of just three feet between photo one and photo two, and aimed a little more to the left. As you can see, the background is virtually totally different. If the mountains were just plain featureless white shapes, as they are in the Apollo photos, the hoax believers would claim that they show two different locations. They don't.

Different backgrounds from the same location

5) 'Wrong' shadows.

This is a general category and covers many photographs based on the shadows being 'wrong'.

This is a good one as well. It shows how easy it is to make wrong assumptions when looking at a 'problem' with tunnel vision instead of trying to understand what is really going on from a scientific point of view. This I feel is a concept that must be alien to Moon hoax believers.

Lots of the hoax claims rest on the belief that the shadows shown in the photographs are somehow wrong, that they indicate more than one light source because the object shown is illuminated from the front and the sides, and so on. This leads them to believe it is due to lighting mistakes on a film set.

The simple fact is that there IS more than one light source. The light does not come directly from the Sun and illuminate only the one object in question, as a narrow beam spotlight would in a dark room. It shines on the entire 'daytime' surface, just as it does here on Earth. Therefore it also illuminates the surface, the astronauts themselves, rocks, mountains, the Lander and all the other objects on the surface. The reflections from these various objects is why there is more than one light source, it is not because there was more than one spotlight used on a film set. It is also worth noting that on the Lunar surface the reflected sunlight from the Earth is 68% brighter than that of the full Moon as seen from Earth.

It is also worth noting that if two spotlights were used this would create two shadows from each object so illuminated. None of the lunar photos shows more than one shadow.

One final point regarding photographs. It has been argued that as all the photographs are so good, they must be fake! Just bear in mind that with over 30,000 taken, NASA was able to select from just the best of them for release. With that many to select from you don't really think NASA are going to include the ones that didn't come out well.

Another question that hoax believers like to keep asking is how was it possible for the astronauts to take such good photographs when they were not even able to hold up the camera and look through the view finder. Practice! Not only lots and lots of practice, but also bear in mind how a poorly framed picture can be made to look good by

Keith Mayes

darkroom cropping. If a photograph shows the subject tilting over at a crazy angle, or way off to one side, good cropping will correct this.

So there you have it, the most commonly believed hoaxes shown not to be.

Personally I find the most surprising thing about the whole business is that hoax believers think that NASA, armed with a budget of billions of dollars and the best experts in the world, could make so many incredibly stupid errors that even the most novice, untrained, inexperienced amateur can spot them easily.

I think that basically, this is what this is all about. Hoax believers consider themselves to be very smart and all the experts at NASA incredibly stupid. This is the same hoax believers that didn't even think to try to photograph the stars themselves, because they just know its a hoax, no need to test it, they saw it on the internet. Oh yes, very smart.

Conditions on the Moon are different to the Earth.

All the other 'fake' photographs are explained just as easily with a little knowledge, and an understanding of how conditions on the Moon are very different to those here. With no atmosphere to scatter the light, things look a little odd on the Moon, it has a very black sky and a very bright surface. We see strong shadows everywhere, and our sense of distance is also fooled because there is no atmosphere to produce the familiar atmospheric haze that creates a distance perspective on Earth. Furthermore, with the gravity being only a sixth of Earth's gravity, things move and behave differently as well. It's hard to make straight comparisons, because we cannot, the Moon is just not like the Earth. We have to think differently when interpreting the images from the Moon, and that's what causes the problems, people are not allowing for those differences when looking at the Lunar photographs. They are looking at them as if they were taken under normal Earth conditions, and concluding wrongly that there must be something wrong with the photographs. There isn't!

Scientific Errors

It isn't just the photographs that have misled the hoax believers either, it's also a lack of scientific knowledge. It's difficult to select a favourite hoax that makes the most ridiculous claim, because there are so many to choose from, but personally I think this has to be number one, its a beauty! I just love it to pieces.

The Moon rocks are just Earth rocks.

You could, I suppose, argue that every laboratory, university, research centre, geological institution, professional scientist, etc, throughout the world that has examined the Moon rocks, are so incredibly stupid that they have failed to spot that they are really only faked up Earth rocks? (This is ignoring the fact that they cannot be faked anyway). Or perhaps you prefer the good old standby that it is all part of a world wide conspiracy. (No, you're not paranoid, they really are all conspiring against you).

Or perhaps instead you think that a probe was sent to the Moon to bring back the Moon rocks? Well yes, that is possible, sort of, but not on this scale. Three robotic Soviet Lunar probes returned a total of about 3/4 lb. (301 grams) from three lunar sites in the 1970's. However, the Apollo crews from 1969 to 1972 collected a total 840 lbs, (382 kgs.) of rock and other surface material. One rock alone weighed 25 lbs. (11.7 kgs.) In comparison to the Apollo total of 840 lbs. the Soviet total of 3/4 lbs. is miniscule. Probes simply could not have returned that much material, (especially a single rock weighing 25 lbs.) and if they could have, it would have been the Soviets that achieved it as they were always way ahead in the field of robotic probes.

Here's the explanation of why they can't be Earth rocks. Extract taken from "The Great Moon Hoax" web site.

"Moon rocks are absolutely unique," says Dr. David McKay, Chief Scientist for Planetary Science and Exploration at NASA's Johnson Space Centre (JSC). McKay is a member of the group that oversees the Lunar Sample Laboratory Facility at JSC where most of the Moon rocks are stored. "They differ from Earth rocks in many respects," he added. Just as meteoroids constantly bombard the Moon so do cosmic rays, and they leave their fingerprints on Moon rocks, too. "There are isotopes in Moon rocks, isotopes we don't normally find on Earth, that were created by nuclear reactions with the highest energy cosmic rays," says McKay. Earth is spared from such radiation by our protective atmosphere and magnetosphere.

Even if scientists wanted to make something like a Moon rock by, say, bombarding an Earth rock with high energy atomic nuclei, they couldn't. Earth's most powerful particle accelerators can't energise particles to match the most potent cosmic rays, which are themselves accelerated in supernova blast waves and in the violent cores of galaxies. Indeed, says McKay, faking a Moon rock well enough to hoodwink an international army of scientists might be more difficult than the Manhattan Project. "It would be easier to just go to the Moon and get one."

"I have here in my office a 10 foot high stack of scientific books full of papers about the Apollo Moon rocks," added McKay. "Researchers in thousands of labs have examined Apollo Moon samples — not a single paper challenges their origin! And these aren't all NASA employees, either. We've loaned samples to scientists in dozens of countries (who have no reason to cooperate in any hoax)."

Astronauts could not survive passage through the Van Allen radiation belt.

Another popular hoax theory, this time sounding more plausible than usual because it is difficult to verify without a very good understanding of the nature of particles and their effect on the human body.

The Moon hoax believers believe that in order to survive the astronauts would have required lead shielding ranging from 4 to 6 feet thick. If that were the case, then the Russians would have discovered this and used it as an excuse for discontinuing manned exploration of space, but they didn't, and have never claimed it to be danger. Actual measurements taken from the Apollo missions show the radiation dosage for the astronauts to be between 1-2 rem, the equivalent of a couple of chest X-rays.

The "six feet of lead" statistic appears in many conspiracist charges, but no one has yet owned up to being the definitive source of that figure. In fact, six feet (2 metres) of lead would probably shield against a very large atomic explosion, far in excess of the normal radiation encountered in space or in the Van Allen belts.

While such drastic measures are needed to shield against intense, high frequency electromagnetic radiation, that is not the nature of the radiation in the Van Allen belts. In fact, because the Van Allen belts are composed of high energy protons and high energy electrons, metal shielding is actually counterproductive because of the Bremsstrahlung that would be induced.

Metals can be used to shield against particle radiation, but they are not the ideal substance. Polyethylene is the choice of particle shielding today, and various substances were available to the Apollo engineers to absorb Van Allen radiation. The fibrous insulation between the inner and outer hulls of the command module was likely the most effective form of radiation shielding. When metals must be used in spacecraft (e.g. for structural strength) then a lighter metal such as aluminium is better than heavier metals such as steel or lead. The lower the atomic number, the less Bremsstrahlung.

The notion that only vast amounts of a very heavy metal could shield against Van Allen belt radiation is a good indicator of how poorly though out the conspiracist radiation case is. What the conspiracists say is the only way of shielding against the Van Allen belt radiation turns out to be the worst way to attempt to do it!

The spacesuits cannot operate their air-conditioning in a vacuum.

It seems to be a popular misconception that heat cannot be removed from an object in a vacuum. A little thought on the subject easily dispels this myth, for if that were the case how could we receive heat from the Sun? Radiation travels though a vacuum, this is an obvious fact. The system used by the spacesuits to shed surplus heat generated by the exertions of the astronauts is fairly simple in operation. The astronauts wear a garment that has water filled tubes running through it that is circulated around their body. The cool water absorbs the heat from the astronauts and is cooled down by the cooling unit. The cooling is achieved by spraying a fine jet of water over the cooling tubes which are located outside the space suit in the cooling unit and exposed to the vacuum of space. In a vacuum the water spray naturally rapidly expands, and in expanding is naturally rapidly cooled, and turns to ice on the tubes. The water passing through is cooled by the cold tubes and the heat given up by the water melts the ice and is lost into space as it evaporates. Simple. The astronauts are able to adjust the temperature control of their spacesuits according to their exertions.

The amount of water used is minute compared to the volume of drinking water taken on board to supply a crew of three for periods of a week and over.

Why doesn't the Hubble Space Telescope provide proof?

This argument runs along the lines that as the HST can provide images of galaxies millions of light years away, why can't it provide images of a Lander on the Moon, which is on our door step?

Bit of a funny question really, anyone with normal eyesight can see the Andromeda Spiral Galaxy easily with the naked eye, and that's over 2 million light years away, yet cannot see a Lander on the Moon! As an amateur astronomer of some 40 years standing I have always understood why the HST could not provide images of the lunar landers on the surface of the Moon, but to get the correct figures I checked out the Hubble Space Telescope web site. Its all down to the size of Hubble's main mirror, which is 2.4 metres. One of the factors of the worth of a telescope is its resolution, the smallest amount of detail it can see, and this depends on the size and quality of the mirror. Hubble's resolution is an amazing 0.048 arc seconds. This is how I calculate the minimum size object that HST can image on the Moon, in as simple a way as I could devise.

HST resolution = 0.048 arc seconds (formula for this is 116 divided by aperture in mm. = 116 divided by 2400)

Visual maximum diameter of full Moon = 31'40" = 1900 arc seconds (a fraction over 1/2 a degree)

Therefore HST can resolve an object on the Moon of (1900 divided by 0.048) = 1/39,583 of the Moon's diameter

Actual diameter of Moon = 3476 km

Therefore resolvable object size = 3476 km divided by 39,583 = 87 metres

As the Landers are less than 10 metres across it is not possible for the HST to resolve them, they just wouldn't show up on any image of the area under examination. I emailed the HST site to make sure I had got my sums right, explaining why I needed it for this book, and their reply was as follows:

"You are correct. Hubble's resolution is good and can resolve objects and areas as small as 280 feet, (86 metres) which rules out the Apollo debris on the moon. Hope this helps!"

Yes it does! Thanks to the HST Office of Public Outreach.

PS. The current largest ground based telescope is the 10 metre Keck, far bigger than the HST and therefore has a far better resolution of 0.012. But this is a theoretical limit that cannot be achieved through an atmosphere, so the HST, being in the vacuum of space, is still number one.

Why haven't we been back?

This one comes up on a fairly regular basis and is used by hoax believers to support their argument that we never went in the first place, because if we had then surely we would have gone back. There are a number of reasons why this has not happened, and it is necessary to know the reason for going in the first place, and the history behind it, to understand why.

On the 25th May 1961 President John F Kennedy told Congress: "I believe that this nation should commit itself, before this decade is out, to the goal of landing a man on the Moon and returning him safely to Earth."

The only reason for making this declaration was in response to the USSR for having put the first man in orbit, because at the time of Kennedy's speech, the USA had only managed one small sub-orbital manned flight. The Moon landing project was not a scientific endeavour, it was a political decision to win the 'Space Race', as it was believed that whoever 'controlled' space would gain an enormous military advantage. It also had very important propaganda value regarding Communism v Capitalism. The USA felt that it was extremely important that they overtake the Russians in the 'Space Race' and be the first to land a man on the Moon. This was a propaganda war at the height of the cold war. Nothing to do with science. See 'The journey to the Moon'

Having achieved the goal of landing a man on the Moon in 1969, that was it, mission accomplished. The Russians had given up trying and a spirit of cooperation was beginning to emerge. President Nixon cancelled the Apollo project, and the last to go was Apollo 17. It had been planned to send Apollo's 18, 19 & 20, but the USA had other far more pressing issues, such as the Vietnam war for example. The American public had become bored with the Moon landings and felt it was becoming a huge waste of public money, and in response to the general apathy many TV channels did not even bother to give the Moon flights air time. (Remember the film 'Apollo 13'). Furthermore, photographs of American soldiers ducking bullets in a muddy trench in Vietnam while listening to the Apollo 11 astronauts walk on the Moon, was, to say the least, incongruous.

There is really not much point in going back, unmanned probes can do the job a lot better, faster, safer and cheaper. Why risk lives? NASA's budget today is invested in numerous projects, such as the Hubble Space Telescope and Shuttle flights to service it. The International Space Station, again serviced by the Shuttle, various probes to study the Sun, Mars, Saturn and other planets, comets and asteroids, and so on.

Going back to the Moon would be unbelievably expensive, and very little would be gained by it. Would the American public readily part with their tax dollars for such a pointless venture when they have other issues, such as health and welfare, unemployment, areas of poverty, a stock market collapse, an energy problem, pollution, crime, etc. (as do most countries I would add) not to mention a very costly war against the 'axis of terrorism'? What President is going to propose a massive investment in returning to the Moon, for no real reason, when there are so many more important issues that need addressing? It would be madness.

That's why we have not gone back. It's not because we have not been there, but because we have.

The last words from the Moon.

Gene Cernan, commander of Apollo 17, said the last words spoken on the surface of the Moon: "This is Gene and I'm on the surface and as I take man's last step from the surface, back home for some time to come - but we believe not too long into the future - I'd like to just say what I believe history will record. That America's challenge of today has forged man's destiny of tomorrow, and as we leave the Moon at Taurus-Littrow, we leave as we came and god willing as we shall return, with peace and hope for all mankind. Godspeed the crew of Apollo 17." (December 1972.)

It's taking us far longer to go back then we ever imagined.

Moon hoax believers - FAQ's The top 20 topics (in no particular order)

I get lots of emails from moon hoax believers asking me questions about the Moon landings. Here are the 20 most frequently raised topics, together with the explanation.

Q1. *Why are there no stars in the sky in the photographs taken from the lunar surface?*
A.   Because the stars are too dim compared to the brightness of the Lunar surface and objects on it.

Q2. *Why does the flag wave in the breeze if it's supposed to be in a vacuum?*
A.   Its not waving. Its held out by a rod running along the top of the flag. It only waves when the astronauts are moving the pole about.

Q3 *Why can't the Landers be seen by the Hubble Space telescope?*
A.   Because they are too small to resolve, even for the Hubble Space Telescope.

Q4 *Why doesn't the Lunar Module show a jet flame when it takes off from the surface?*
A.   Because the type of fuel used by the Ascent Stage doesn't show in a vacuum.

Q5. *The astronauts couldn't have controlled the lander during descent because the centre of gravity would change every time they moved.*
A.   The descent is kept level by small computer controlled thrusters.

Q6. *They could not have survived travelling through the Van Allen Belt without suffering from radiation sickness, or death, without a 6 feet thick solid lead shield.*
A.   They did survive. Sources that claim the radiation would have been lethal use creative and incorrect figures. Thin metal provides better protection than thick metal.

212

Q7. *NASA killed the astronauts that were going to reveal the hoax.*
A. Malicious and evil nonsense.

Q8. *The Landers should have kicked up a lot of dust when landing, but there is no sign of any dust on the landers in the photographs.*
A. The landers were throttling back when landing and using little thrust. Also they didn't come straight down but at an angle travelling over the surface. Touch down kicked up only a little dust that in a vacuum quickly fell back to the surface.

Q9. *The astronauts would not have been able to make footprints on the Lunar surface as the dust would be too dry.*
A. The surface is dry, because the dust is in a moisture free vacuum. Therefore the particles are not worn smooth due to weather erosion, but are gritty and angular, thus tending to lock together under pressure. There is also a more technical reason too complicated to summarise here.

Q10. *The cross hairs on the photographs have been pasted on afterwards and go behind some objects. (*Pasted on afterwards after and go behind......??? No, never mind)
A. The crosses are sometimes rendered invisible when shown against a bright background, this effect is known as bleed over. If you check you will not see this effect against a normal or dark background.

Q11. *The Landers should have sunk deep into the lunar dust, but they didn't.*
A. Why should they? The dust is only shallow.

Q12. *The video shows Neil Armstrong climbing down the ladder and stepping onto the surface. If he was supposed to be the first man on the Moon, who took the video?*

A. A video was deployed by Armstrong on an extending arm swung out from the Lander for this purpose. Too big an occasion to miss.

Q13. *Heat cannot escape in a vacuum, the astronauts would have fried to death in the sun.*
A. The only heat the astronauts could receive in a vacuum is by radiation. Their suits were very reflective and internally cooled.

Q14. *Heat cannot travel through a vacuum, they would have frozen to death going into a shadow.*
A. Again, because they are in a vacuum, the only way they could loose heat is by radiation. Their suits were insulated.

Q15. *The Moon rocks are just faked up Earth rocks.*
A. Not possible. Rocks that have aged in a moisture free vacuum, exposed to cosmic rays and meteorite impacts, cannot be faked from Earth rocks, far too different. They have been examined by labs. throughout the world and all agree they are Moon rocks.

Q16. *Some of the photographs show shadows from more than one light source, proving it was all shot in a film studio in the Nevada desert using multiple studio lights.*
A. Sunlight on the atmosphere free moon is very bright, and is reflected off every object on the surface. Even the surface itself reflects a great deal of light into areas that would otherwise be in shadow. So yes, there is more than one light source.

Q17. *They didn't have the computer technology in those days to get to the Moon and back.*
A. Most of the computing was carried out on huge main frames by NASA and the required data was uploaded to the small onboard computer. Computer technology was already advanced enough for the country's nuclear defence missile system.

Q18. *If there were only two astronauts on the surface at any one time, who took the photographs, because none of the astronauts are seen holding a camera?*

A. The cameras were strapped to their chests, they couldn't easily hold them in their gloved hands and carry the various instruments around. Furthermore, they could not hold them up to their eye anyway to look through the viewfinder because of their helmets.

Q19. *One of the Moon rocks in the photo has the letter 'C' written on it, clearly showing it to be a film set prop.*

A. The letter 'C' is not on the original print. It only appears on duplicates and is nothing more than a piece of fluff or hair that got into the equipment.

Q20. *The camera and films used could not have survived the extremes of temperature on the Moon*

A. The fact is they did. Check out Hassleblad's web site at http://www.hasselblad.com and Kodak's site at http://wwwuk.kodak.com for full technical details.

The journey to the Moon

On the 25th May 1961 President John F Kennedy told Congress: "I believe that this nation should commit itself, before this decade is out, to the goal of landing a man on the Moon and returning him safely to Earth."

Many people have expressed their amazement that not only was the goal of landing a man on the Moon achieved, but that it was achieved in only 8 years, as Kennedy said it would. This is however, ignoring the fact that at the time Kennedy made his statement NASA already had in the pipeline over nine different Moon landing flight plans in a project they had named 'Apollo'. They were already designing a huge Moon booster called 'Nova', that was to generate 40 million pounds of thrust, and were already considering various methods for landing a man on the Moon. But at this stage that information was not yet in the public domain.

It was not a coincidence that man landed on the Moon in the last few months of the decade and thus achieved the goal, it was because NASA used all the time available to thoroughly test, check and triple check every single step along the way. The Moon landing, if they had been pushed, could have easily been done with a few amendments by Apollo 10, Apollo 9, or even before that by Apollo 8 in December 1968.

Had NASA not been put under pressure to meet Kennedy's deadline, they would have chosen a far different approach to land a man on the Moon than the one used. It was originally hoped to do it stage by stage using a permanent Earth orbiting station that would make future flights a lot easier, but instead had to settle for a 'one time' system to meet the deadline. With the new system going from launch pad, to orbit, to the Moon and back, using disposable components, it was possible to achieve within the time frame, but it meant each mission was a 'one off' and contributed nothing towards the overall mission plan that could be used by following Moon flights. It meant that with every flight NASA had to throw away a Saturn V rocket costing $185,000,000!

The mission to land a man on the Moon was not an 8 year period of starting spaceflight from scratch and ending with a Moon landing. Spaceflight began in 1957 with the first satellite, Sputnik 1, placed in orbit by the USSR, and developed from there. The first American satellite to reach orbit was Explorer 1 in January 1958.

The launch of the first satellites spurred rocket scientists into action, and with only five satellites safely in orbit the first attempt was made to send a spacecraft to the Moon.

August 1958 saw the launch, by the USA, of the first unmanned rocket to attempt to reach the Moon. It failed, as did the Russian rocket launched a month later. By December 1960 The USA had made a total of nine Moon attempts with one successful fly-by. The USSR had made seven attempts with one direct hit and one Lunar loop.

So far, not very encouraging, but at least unmanned rockets had reached the Moon. During this period both countries were sending animals into orbit to pave the way for manned flights.

MERCURY

NASA was formed on October 1st 1958, and the man in space programme was introduced just six days later, almost three years before Kennedy's pledge to land a man on the Moon. The program was renamed "Project Mercury" on Nov. 26, 1958, just prior to the commencement of the astronaut candidate selection process. NASA selected seven pilots to train for flights in the one-man capsule called Mercury. It was a bell-shaped capsule that could be controlled in space by its pilot, manoeuvring in three axis called pitch, roll and yaw. The pilot could take full manual control or just monitor automatic systems. He had the ability to override systems and troubleshoot problems. The Mercury had an ablative heatshield on its blunt end which would take the brunt of the intense heat on its high speed re-entry into the Earth's atmosphere, carefully controlled at a specific angle. A parachute would enable the craft to splash down in the sea. The first manned flights would involve sub orbital "up and down" rides launched on a Redstone rocket and would be followed by orbital flights on the Atlas, America's first ICBM.

A major American milestone was reached with a Redstone boosted sub orbital flight of the chimpanzee Ham, in January 1961. This was soon put into the shade however by the USSR three months later with the first man in orbit, Yuri Gagarin. The American response was their first manned spaceflight, using a Mercury capsule, Freedom 7, a brief 15 minute sub orbital flight on May 5th 1961. No match for the Russian Yuri Gagarin's trip into orbit. The Soviet premier Nikita Khrushchev was ecstatic and milked all the propaganda he could from the flight. This did not go unnoticed by America's newly elected President and 20 days after Alan Shepherd's space hop, Kennedy responded to the Soviet lead by making his pledge to land a man on the Moon before the decade was out. The race was on.

COUNTDOWN: 8 years 7 months remaining.

NASA had already studied three options of landing a man on the Moon.

1) The Direct Ascent Method. This would involve the construction of a huge booster, the Nova, that would launch a large spacecraft and send it on a course directly to the Moon. The craft would land on the Moon, and after a period of exploration, would take-off and fly directly back to the Earth. This method was ruled out as being too expensive and requiring too high a level of technical sophistication of the Nova.

2) Earth-Orbit Rendezvous. This called for the launching of all the components required for the Moon trip into Earth orbit, where they would rendezvous, be assembled, refuelled, and sent to the Moon. This method was dropped due to problems associated with manoeuvring at rendezvous and assembling components, and dangers of refuelling.

3) Lunar-Orbit Rendezvous. This proposed sending the entire lunar spacecraft up in one launch. It would head to the Moon, enter into its orbit, and dispatch a small lander to the lunar surface. It was the simpler of the three models, but it was risky. In the plan, three astronauts would be launched in a mother ship, first reaching Earth orbit, then heading for the Moon, to enter an orbit around it. A landing vehicle, manned by two astronauts, leaving one in the mother ship, would touch down on the Moon using its descent engine. After a Moonwalk, the top half of the Lunar Excursion Module would take off, leaving the bottom half on the surface, and rendezvous and dock with the mother ship in lunar orbit. The mother ship would break out of lunar orbit and head back to Earth. Since rendezvous was taking place in lunar, instead of Earth orbit, there was no room for error or the crew could not get home. Also, some of the most difficult course corrections and manoeuvres had to be done after the spacecraft had been committed to a circumlunar flight.

This method, though risky, was adopted in 1962 as it was technically the simplest, and the Apollo project was on its way.

COUNTDOWN: 8 years remaining.

By now the Americans were having great success with their Mercury programme. On the 20th February 1962 John Glenn was launched into orbit by an Atlas rocket in a Mercury capsule called Friendship 7. The first American in orbit, he completed 3 orbits. This was followed by Scott Carpenter with 3 orbits, Wally Schirra with 6, and Gordon Cooper with 22.

COUNTDOWN: 7 years 10 months remaining.

Rendezvous and docking of spacecraft together was to be a crucial part in the Lunar-Orbit Rendezvous method and the next series of US piloted spacecraft that succeeded Mercury was designed to demonstrate these manoeuvres in Earth orbit and to rehearse as much Moon flight as possible without going there. Testing out spacesuits during spacewalks (EVA's) and flying in Earth orbit longer than it would take to fly to the Moon and back were also on the agenda for this next series of spacecraft which would carry two astronauts.

Mercury met all three of its objectives: orbit a manned spacecraft around Earth; learn about man's ability to function in space; and safely recover the man and spacecraft. The project ultimately put six men in space, four of whom made orbital flights around Earth. It proved that men could function normally for up to 34 hours of weightless flight. Over two million people worked on the project for almost five years. By 1963, Project Mercury wrapped up and Project Gemini was two years into its development stages.

COUNTDOWN: 6 years remaining.

GEMINI

The Gemini spacecraft would become the first to alter its orbit and manoeuvre in space, which was crucial for it to be able to rendezvous and dock. Two unmanned test flights of Gemini were made before Gemini 3 made its first manned flight in March 1965. It carried two astronauts and made a modest three orbits, but successfully demonstrated the first manned manoeuvres in orbit as a crucial test for Apollo. The command pilot was Gus Grissom, who had flown the second sub orbital Mercury mission in July 1961 and the first person to make two spaceflights.

COUNTDOWN: 4 years 9 months remaining.

The remarkable Gemini programme then soared ahead with nine more piloted flights ending in 1966, meeting all its goals during one of the most frenetic and exciting periods of the Moon Race. Astronauts Neil Armstrong and David Scott completed the first space docking on 16 March 1966 when Gemini 8 joined up with an unmanned Agena target rocket, simulating the ascent of a lunar module from the Moon, docking with a mother ship in lunar orbit.

COUNTDOWN: 3 years 9 months remaining

Meanwhile in June 1966, when Gemini 9 was flying, the first mock-up of a Saturn V booster was being rolled out at Kennedy Space centre.

COUNTDOWN: 3 years 6 month remaining

The remarkably successful Gemini programme ended with the landing of Gemini 12 in November 1966 and NASA felt confident that the major requirements for a Moon mission had been mastered.

COUNTDOWN: 3 years 1 month remaining

MOON PROBES

From the period 1962 onwards, both America and the USSR were involved in sending unmanned probes to the Moon with the aim of finding suitable landing sites. Some scientists, for example, thought that a craft would disappear in a vast layer of soft lunar dust. Fortunately this turned out not to be the case. America launched its Ranger series of probes in 1962, and in 1965 Ranger 8 and Ranger 9 returned over 12,000 images before crashing into the surface as designed. Ranger 8 imaged the Sea of Tranquillity, and this was eventually selected as the first landing spot. This was followed in 1966 by Surveyor 1, a soft landing probe, that sent back spectacular images from the surface that were shown live on TV. In 1967 Surveyor 3 soft landed and scooped up surface material. Lunar Orbiter had 5 successful missions and sent back thousands of pictures, almost the entire Moon. This enabled NASA to select up to 20 candidate landing sites for the Apollo programme.

APOLLO

The Apollo lunar spacecraft comprised of three major components, the Command Module, the Service Module and the Lunar Module. The crew flew to and from the Moon in the Command Module; the Service Module was attached at all times to the Command Module until just before re-entry into Earth's atmosphere, when it was jettisoned. The Lunar Module was the compartment in which two of the crew landed on the Moon and took-off again to join the Command Module, which remained in lunar orbit. Another major element of the spacecraft during the first 100 seconds of flight was the launch escape system in case the Saturn V booster malfunctioned. At the nose of the Command Module was the docking mechanism that allowed it to join up with the Lunar Module that was in fact beneath it for the launch. Once dispatched towards the Moon, the Command Module and Service Module combination separated from the third stage of the Saturn V rocket, turned around and docked with the Lunar Module nestled inside the third stage and extracted it from the spent booster.

The combined craft flew to the Moon, with the crew able to transfer to and from the inhabitable modules via a transfer tunnel, once the docking probe had been removed. A vital part of the Command Module was the heatshield which protected the crew from the 1600 C (3000 F) temperatures experienced during the plunge into Earth's atmosphere, which begins at 25,000 mph.

THE APOLLO 1 DISASTER
The intense efforts being made to get the whole Apollo system in gear for the first piloted flights was illustrated by the development of the first Saturn V to its first launch. It took just five years.

It has to be remembered that an awful price was paid. On Friday 27th January 1967, Apollo 1 was on launch Pad 34 at Cape Canaveral on top of the massive Saturn V booster, for what was to be a countdown demonstration test during which the rocket was unfuelled. This was to be followed on 14th February 1967 with a shakedown orbital flight. There were three crew onboard, Gus Grissom, Edward White and Roger Chaffee. The Command Module, as usual, was pressurised with pure oxygen. Bad workmanship had resulted in some electrical wiring losing its insulation and this caused a spark under Grissom's seat. Within seconds the arc had become an inferno in the oxygen atmosphere. All three men died within 15 seconds. Those men were test pilots, but more, they were heroes of spaceflight.

The Apollo 1 disaster revealed carelessness and bad workmanship in design and production. The programme was delayed while modifications were made.

COUNTDOWN: 2 years 11 months remaining

SATURN V
On 9th November 1967, the first Saturn V booster was launched from Pad 39A at the Kennedy Space Centre. With the Apollo 4 system on top, the monster rocket was 363 feet high and weighed 2,888 tons.

Versions of the Apollo modules had made a number of previous test flights in Earth orbit using smaller Saturn 1 and 1B boosters. The Lunar Module was due for testing in early 1968.

COUNTDOWN: 2 years 1 month remaining

In 1968, NASA planned one Earth orbit mission, to be followed by a combined Apollo Command and Service

Module and Lunar Module test mission in Earth orbit, after launch on a Saturn V. This would be followed in 1969 by a deep Earth orbit test and a final Moon landing dress rehearsal in Lunar orbit. If all went well an American could be on the Moon by mid 1969.

Apollo 5
January 22 1968 1st test of Lunar Module in space
COUNTDOWN: 1 year 11 months remaining

Apollo 6
April 4 1968. Final uncrewed Apollo test flight. Full systems check.
COUNTDOWN: 1 year 8 months remaining

Apollo 7
October 11 1968. First manned Apollo flight. Earth orbit and test of CSM
COUNTDOWN: 1 year 2 months remaining

Apollo 8
December 21 1968. This was the first mission to place men into an orbit around the Moon. It was hailed a huge success. The Moon landing was becoming a reality.
COUNTDOWN: 12 months remaining

Apollo 9
March 3 1969. This flight followed Apollo 8 but stayed in Earth orbit. It then had for the first time two crewmen entering the Lunar Module and separating from the Command and Service Module. After a brief test flight and testing of the descent engine it re-docked with the CSM. Only the actual descent to the lunar surface remained to be tested. Everything else was in place.
COUNTDOWN: 9 months remaining

Apollo 10
May 18 1969. This was a full rehearsal of a Moon landing, with the Lunar Module making a descent from lunar

orbit to within 9 miles of the lunar surface before firing its engine and returning to dock with the Command and Service Module. Every system, every procedure, to be used in the actual Moon landing was tested, apart from the actual landing itself, and worked flawlessly. They were now ready to make the attempt to land on the Moon.

COUNTDOWN: 7 months remaining

Apollo 11

Apollo 11 blasted off from the Kennedy Space Centre on 16 July 1969, watched by one million spectators from the nearby beaches and causeways, and 600 million people around the world, including me. Neil Armstrong stepped out onto the surface of the Moon on 20th July 1969.

COUNTDOWN: Goal achieved with 5 months to spare.

The rest is history

Did those twelve Apollo astronauts *really* walk on the Moon? Of course they did! To believe otherwise is an act of sheer stupidity.

Footnote:

Following the Columbia disaster when the Space Shuttle disintegrated over Texas during re-entry on 1st February 2003, questions have been raised over the problems associated with re-entry.

When in orbit the Shuttle is travelling at a speed of 27,000 km/hr (app. 16,800 mph) and this speed has to be shed as it descends through the atmosphere. In the early days of space flight the Mercury capsule descended through the atmosphere much steeper than today's Shuttle, and much faster, with the result that it was much safer. This may sound odd but it's all due to the shockwave created by the space craft, and the blunter the capsule and the faster it moves, the farther away the shock wave occurs. A returning spacecraft does not get hot through friction with the atmosphere. A layer of air builds up in front of it, and between this layer and the surrounding atmosphere is the shockwave. That's where the heat is generated. Most of the heat of a Mercury re-entry was swept away into the atmosphere.

Because the shuttle was required to be manoeuvrable during landing it had to have a delta wing which meant that the wing is much more exposed to the heat. NASA originally wanted the Shuttle to re-enter like the Mercury capsule, with its nose held high, but it's shape and size restricted it with the nose allowed no higher than 40 degrees to the horizon. As a result, the Shuttle would be subject to four times as much heat, for twice as long.

The shuttle re-entry procedure has always been dangerous, as is the lift off, strapped as it is to the External Fuel Tank and two Solid Rocket Boosters, with no means of escape for the crew in the event of an emergency.

NASA, the US Space Agency, has calculated that it would lose one Space Shuttle in 438 missions. The reality is it has lost two in just 113 missions. That's *eight times* worse than predicted.

Time for a change.

19. UFO's and Alien Abductions?

"Nothing in all the world is more dangerous than sincere ignorance and conscientious stupidity."
Martin Luther King Jr.

Judging by the number of web pages devoted to UFO's, let alone magazines, associations, newsletter groups etc, it would appear that there are a great many people who believe in the existence of UFO's. By UFO's I am referring of course not to unidentified flying objects, of which there are many, but to "flying saucers"- spacecraft from another planet. So why is it that so many people believe in them, when the vast majority of us have never seen one? One of the reasons perhaps is because of the huge volume of reported sightings from around the world. With so much 'evidence' it is hard to ignore the possibility that we are being visited by 'flying saucers'.

So what actual proof do we have that UFO's exist? None whatsoever! So why do so many people believe that they have actually seen them, or that others have? Let's examine the 'evidence': Let's first though take into consideration that lots of people claimed to have met Elvis recently, have seen the Loch Ness monster, receive telepathic messages from aliens on Venus and know for a fact that the Moon landings were shot in a film studio. See "Did we land on the Moon?" People are funny. All unusual reports need to be viewed with a fair dollop of caution.

Eye witness sightings.
Many people claim to have seen a UFO. This is not surprising. For over thirty years I have been a keen amateur astronomer and I have spent a great deal of time looking at the sky, both during the day and at night. I have come to know what I am looking at, I have gained a lot of experience of observing objects in the sky and I have not seen a single UFO, but I have seen plenty of things that less experienced observers would classify as a UFO.

People tend to scoff when you tell them that the huge, bright, flickering object they saw hovering near the horizon was not in fact a UFO but Venus. The same goes for Sirius and other very bright stars when seen low in the sky. Under certain conditions they do look like UFO's, as do some cloud formations. Lets also not forget satellites arcing across the sky, and the International Space Station. I was watching the ISS a couple of months back, it appeared from nowhere, arced across a perfectly clear sky for about a minute or so and then just disappeared! (It had in fact entered the Earth's shadow). I wondered at the time how many people witnessing that would believe it to be a UFO accelerating into hyperspace!

Our local police helicopter operating at night with a powerful searchlight and weaving about all over the place and at varying heights is also an impressive candidate. I was once fooled for a time by rotating laser beams reflecting off low level cloud, it was really impressive and had me going for a while. But it was just a disco putting on a laser show. Even meteorites and comets have been mistaken for UFO's, not to mention 747's with their landing lights and strobes illuminating low level clouds. I could go on but I won't. You see my point, being human we are prone to error, and sometimes we see what we want to see.

Quite obviously the vast majority of so called UFO's can be explained away by natural phenomena, but that still leaves a small percentage that can't be explained away, and it is this small percentage that ufologists believe may be 'flying saucers'. However, because an object is classified as 'unidentified' it does not automatically become a spacecraft from another planet. That is wild speculation. Why not just accept the obvious, that although we can't be sure what it is, because of poor video or photographic quality, conflicting eyewitness accounts, vague descriptions etc. making identification impossible, it does not mean that it is an alien spacecraft. It could be a secret test flight of an advanced fighter jet, or just a hot air balloon.

227

Photographic and video 'evidence'

I have yet to see a convincing video clip of a UFO. Most of them are downright ludicrous. I have even seen a video that claimed to be of a flying saucer that was nothing more than an out of focus shot of Venus through a wobbly hand held video filming through a window. (I have looked at Venus through my own camera enough times to recognise it as easily and instantly as you would recognise your own face in a mirror). Anyone who has tried to video - or photograph - a dark sky at night through a glass window from an illuminated room will know that the automatic setting on the focusing goes crazy, It 'hunts' back and forth and causes the star or planet to 'pulsate', add on hand wobble, multiplied by the magnification factor, and there you have your classic, flickering UFO, zooming about the sky performing impossible manoeuvres. Really impressive, but it's just Venus and it's not doing anything.

The same goes for the planet Jupiter, if I had a dollar for every video clip of Jupiter that was held to be a UFO, I would be a very rich man. If you want to watch videos of Jupiter that claim to be UFO's there are lots of them on the Net, just go to any search engine and type in "UFO". Some are so bad they are really funny, well they would be funny if it didn't take 10 minutes to download the rubbish. On one particular site I was able to immediately identify, Jupiter, a lens flare from a street light and a lens flare from the Moon, (all fantastic shots of UFO's) and was subsequently proved correct when the hoax was revealed. Unfortunately, a great many other people who visited this site were convinced (and no doubt still are knowing the way they yearn to see UFO's) that they were genuine UFO's, and will point to this as yet more 'evidence'. Going back to 'how can you ignore so much evidence?' Easy, an ounce of knowledge and common sense is worth a ton of extreme speculation and wishful thinking.

As for other videos, they are generally either out of focus, nothing to judge scale and distance by, too much camera shake, poor lighting, it could be just about anything, and so on. The only really good shots are obvious fakes, even obvious to ufologists. And that's another problem, it's just too easy to fake. No video or photograph can be accepted as proof simply because it isn't proof, it could have been faked. As for people that present fakes as the genuine article, they muddy up the water and give serious ufologists a bad name.

So what do we need as hard irrefutable evidence. We need something we can touch! Isn't it strange that with so much UFO activity around the globe for so long, at least 60 years, not one tiny piece of hardware has turned up. Not so much as a single rusty screw from a flying saucer gangway! Not one has landed, or crashed, in an area where it could be 'seen' by more than one or two people. And please do not tell me that there is a complete UFO tucked away in Area 51. Yes, of course there is, we all know that, its been a big secret for years. This from a nation where even its last President, the most powerful man on the planet, was unable to hush up a minor indiscretion involving just one other person! But the entire military personal of an air force base can be persuaded to keep their mouths shut for years. I don't think so.

Motive

If, despite all of the above, you still believe in UFO's, why do you think that the visiting aliens are not revealing themselves? Why do you suppose that they have travelled across light years of space just to skulk around? Or do you think they have made some sort of deal with the government? Why? They hop across light years of space, have mastered gravity, acceleration and inertia, and heaven knows what else, and need to make a deal? And why are they happy to be seen buzzing planes and so forth but not want to contact us? I just can not bring myself to believe that any sufficiently advanced civilisation that can hop between the stars would behave in such a ridiculous manner.

Imagine the day when we finally make it to the stars and find a planet populated by a technologically inferior species. Do you think we would secretly contact the governments of that world and then spend the next sixty odd years hopping about from one place to another, hiding from the general population, while at the same time allowing them to see us coming and going? For what purpose? It would make no sense.

Alien Abduction

These stories tend to fall into two main categories. 1) "They examined my body" 2) "They give me a tour of the spaceship"

1) The 'kidnapped' were strapped/tied down onto a form of operating table and powerless to resist. They were naked and vulnerable. The aliens stare at their exposed bodies and explore them with their hands. Probes are inserted into various orifices. Some even claim the aliens forced sex onto them. The aliens tried to remove this incident from their minds but failed as it all came out under hypnosis. All very surreal. Just the sort of sexual fantasy some people would pay good money for. I believe that's all it is, a fantasy. Maybe for kicks, perhaps for a few minutes of fame, perhaps to make a little money. Maybe they live in a fantasy world, or perhaps it's just to brighten up an otherwise dull, boring and very average life. Reality it isn't. But that's just me expressing my opinion, I could of course be wrong (I don't think so though to be honest). Problem: Why do the aliens do it? They could learn all they need to know from us, or the government they are supposed to be secretly working with, or just some poor old tramp they pick up and then kill. An alien race that is prepared to abduct perfectly innocent people to experiment on would surely not balk at killing them, they obviously see us as nothing more than lab rats. To release these people after bringing them into their ship is far too risky, they must know by now their memory erasing technique is badly flawed.

2) They stumble across a flying saucer, in a dark wood for instance, usually after their car has mysteriously broken down, along with the radio, and taken aboard by the aliens. They are given a guided tour round the ship, see screens showing the stars etc, and sometimes even given a little joy

ride where they look out the window and see the globe beneath them. How nice. The ship lands again. They have the entire episode erased - but not very well - from their minds and are then politely shown the door. All is revealed under hypnosis, just as in case 1). (They still haven't perfected this memory erasing trick yet). Problem: What on Earth is the point of being invited on board, given a guided tour and free joy ride, only to have the entire incident removed from their minds ten seconds later? It really makes a lot a sense doesn't it! As much sense as being invited on board in the first place. This just has to be complete and utter rubbish and as far removed from reality as you can get. It does their case nothing but harm to trot out this sort of obvious nonsense.

Just in case you think these stories must be true because they were 'dug out' of the 'victims' mind while under hypnosis, a study of psychology will explain all. Lots of people have, while under hypnosis, revealed details of their previous lives for example. They were typically Napoleon, Cleopatra, Joan of Arc, etc etc. If you belief abduction stories to be true because of this method, then I am afraid you will have to take on board a whole load of other stuff as well. Most of it even more bizarre then Napoleon being re-incarnated in hundreds of people at the same time! Not to mention those that can 'recall' coming from Pluto. Better just forget it.

If you believe I am wrong, and I know many people do, why then do you think the aliens do it? None of these abductions serve any useful purpose to the aliens whatsoever. They only serve to give us more information about them, which judging by their strange behaviour with the 'flying saucers' appears to be the last thing they want to do.

Give me just one good reason for these so called abductions.

Reasons to believe

Time and again people email me to ask how can I ignore so much 'evidence' for UFO's. My answer is always the same - THERE IS NO EVIDENCE! Typically they will refer to newspaper reports and so called 'sightings'. The following article is a typical example of how the media works. The headline for this story reads "A SUPER HIGHWAY FOR ALIENS? and appeared in a popular UK newspaper, The Daily Mail, on January 15th 2003. The story details how a Spanish businessman downloaded the images, using a satellite dish, direct from the SOHO satellite that is in permanent orbit observing the Sun. He claims to have collected many images of UFO's captured by SOHO. The story goes on to say that these UFO's are seen to change direction and must be controlled by some alien intelligence. It says that NASA was contacted and replied that the images may be due to camera faults or could be comets. UFO supporters claim these images are the most convincing proof yet of UFO's are that NASA refused to answer any more of their questions.

The image of the 'UFO' showed a sharply defined and beautifully coloured saucer shaped object viewed edge on. Below this image, which it is claimed is simply an enhancement, was shown the object as it originally appeared. It was a very small white blob with a straight line going through it, making it look a little like Saturn viewed edge on.

My immediate reaction to the image of the 'UFO' is that it was a very over worked enhancement of the original white blob which is probably just a planet or comet. This is typical of the result that I can achieve very easily in Photoshop if I over-work a planet or galaxy, it is possible to create just about any effect you want. The result achieved here, transforming that fuzzy white blob into the clearly defined, and beautifully coloured finished image, is about as fake as it gets. The finished 'enhanced' image is nothing like the image captured by SOHO! Very often, a bright light source, such as a star or planet, will show 'spikes' as seen here, but in reality the image itself is just a dot.

I contacted the people at SOHO and asked for their views on the matter. Here is their reply:

>>>>Subj: Re: DR.SOHO:

Date: Tuesday, January 21, 2003 4:53:28 pm

From: thompson@orpheus.nascom.nasa.gov

To: Keithmayes123@aol.com

cc: drsoho@sohops.gsfc.nasa.gov

Keith:

We also tend to find these claims amusing, although we find it a bit sad. Actually, although most of the SOHO imaging instruments operate in the ultraviolet and extreme ultraviolet, to observe the Sun in a wavelength regime that can't be seen from the ground, there are a couple of SOHO instruments which do operate in the visible. Most of the UFO claims are based on visible light images taken by an instrument called LASCO, which takes advantage of the airless conditions of space to observe the faint corona around the Sun. The other visible light instrument is MDI, which is studying the Sun through helioseismology, but I've never heard anyone claim to have seen UFOs in MDI images.

People also claim to see UFOs in EIT images, which are taken in extreme ultraviolet light. However, just because an object is visible in extreme ultraviolet doesn't mean that it couldn't also be seen by the eye. After all, the Sun glows in both wavelength regimes.

I completely agree with your description of the images being overworked. The people making these claims should at least have the courage to show the actual data, and not something which has been manipulated in Photoshop. The supposed UFO images that we've investigated tend to fall into several categories:

Planets: These always look very strange in LASCO images, because they're so bright that the image blooms, and the CCD pixels bleed along the readout rows. Some people try to claim that they're flying saucers, based on their appearance. I've also heard the claim that they're previously unknown Saturn-like planets with rings around them. You can see what I'm talking about at http://sohowww.nascom.nasa.gov/hotshots/2000_05_03/

Cosmic rays: High energy particles from the solar wind, and from the galaxy as a whole, whip around the SOHO spacecraft and interact with the detectors. These produce spots and streaks on the detector ranging from a single pixel, to large streaks that span a large fraction of the image. These are most evident during a solar storm, as can be seen in at http://sohowww.nascom.nasa.gov/hotshots/2001_11_26/ but are always present at some level. I know that some people have claimed that they've seen spacecraft-looking things that seem to be moving around, but which are obviously cosmic rays when examined by an experienced observer. People see a cosmic ray at one location in one image, and then another random cosmic ray hit nearby in the next image, and claim they're both the same thing moving between frames. Sometimes you'll see a cosmic ray seem to persist in the web images for two or more frames. This is because we lose a certain percentage of the data coming down from the spacecraft. In LASCO such losses appear as square blocks in the image. The software which puts the images up on the web will fill in these blocks from the last good image, and if there's a cosmic ray in that block from the previous image, it will appear in this image as well. The way to check for this is to look at the raw data files, which are also available on the web through the SOHO catalogue interface at http://sohowww.nascom.nasa.gov/data/

Software glitches: Occasionally we'll have some problems with the software which produce the images for the web, and strange artefacts will appear in the data. These glitches are usually corrected within a few days. In fact, we had a couple of instances of that recently.

Detector defects: There are defects which appear in the cameras from time to time, sometimes temporary and sometimes permanent. I remember seeing a web site which claimed that strange lights were hovering over the lower left limb of the Sun in EIT images, and thought to myself "You only just noticed that?". Those defects have been around forever, and were seen in the lab even before SOHO was launched.

That all said, it should be noted that we do see objects moving in SOHO images. Over 500 comets have been discovered in SOHO images, most by amateurs using LASCO data which have been downloaded from the web. That's more comets than from any other observatory, either from the ground or in space. People are looking for moving objects in these pictures all the time, and are highly motivated to find them. None of them have ever turned out to be anything other than comets.

William Thompson<<<

So there you have it, no UFO's. The thing to bear in mind is that as far as the general public are concerned, they have seen headlines, and images, claiming the existence of 'amazing evidence' of UFO's, and that NASA is refusing to answer questions on it. You can see what nonsense that is. The people at NASA, after politely answering all the questions put to them about the UFO nonsense, eventually have nothing more they can add. There is no conspiracy of silence, so beloved by the UFO brigade. The media, being in the business of selling stories, see the potential for using such wild headlines, it sells a lot of copy, but they don't print the official version next day explaining why it is all a load of nonsense, that doesn't sell newspapers.

As I keep saying, and will continue to keep saying, THERE IS NO EVIDENCE! Stop believing all the rubbish you read in newspapers, and on the weird and wacky web sites that are so popular on the Net, just try using a little common sense.

I would like to believe they are here, but I don't. Wishful thinking should never be allowed to get in the way of facts. I do believe there is extraterrestrial life out there, see "Is there any other life in the universe?" but I am not convinced that we are being visited by beings from another planet, not now nor in the past. I consider that those people who honestly believe that they have seen UFO's are genuinely mistaken. We have absolutely no proof that these flying objects, although unidentified, are 'flying saucers'. They could be anything.

I am not convinced that we have ever been visited by a single UFO, even though there is so much 'evidence' There is also 'evidence' that we never landed on the Moon and that people can levitate, read minds and move objects by telekinesis, etc. etc. etc. Sure they can.

It simply has to be remembered that, regrettably, the world is populated by a great many people who - quite honestly - make absolutely terrible witnesses; ask any police officer!

20. Did aliens build the pyramids?

"The two most common elements in the universe are Hydrogen and stupidity."
Harlan Ellison

I can remember standing in the desert near Cairo and gazing across the Sphinx to the Great Pyramids of Giza, and in particular to the pyramid of Cheops, also known as Khufu, one of the most impressive structures on Earth, and counting myself extremely fortunate to be there. This pyramid carries some impressive statistics. It stands 137 metres tall (450 feet) and has sides 230 metres long and consists of two million blocks of stone, weighing from the most common at 2.5 tons up to 50 tons. The weight of the pyramid is six million tons. Its base is level to better than 25mm (one inch). It deviates from a perfect square by less than three arcminutes, and the side lengths differ by less than 50mm (two inches). The four corners are almost perfect right angles and align almost exactly to the four cardinal points of the compass. The sides slope at almost 52 degrees, or would do had not the smoothly polished limestone finishing blocks been removed and used in building works in Cairo. It is one one of the original Seven Wonders of the World. Impressive though the statistics are, they do not reveal the true magnificence of this wonderful construction. To have seen this pyramid when perfectly smooth and dazzlingly white in the desert sun must have been an incredible sight.

237

So impressive are the pyramids that it has caused many to wonder how they could possibly have been built some 4500 years ago. Some even go as far as to suggest that it is not possible for these structures to have been made by the labours of men, but were instead made by aliens.

Who built the pyramids, aliens or men?

The case for the Aliens:

The pyramids are so accurately aligned with the points of the compass that only aliens could have achieved this all those thousands of years ago. The angle of the slope of the sides is so precise only aliens could achieve this. The blocks are so heavy and the pyramid so tall only aliens could achieve this. In the period 2500 BC man did not have the tools or knowledge necessary to build the pyramids, so only aliens could have done it. How the aliens built the pyramids is not known, but they would have employed the use of advanced construction equipment. The knowledge and skills required to build such vast pyramids to such precise accuracy was beyond human capability.

The case for men:

Never underestimate the ingenuity of man. We are today so used to using machinery to carry out virtually all our major construction work that we sometimes forget that machinery, in terms of historical events, is a very new development, its only been around a couple of hundred years or so. Mankind managed very well without it for many thousands of years. We have long forgotten the techniques that were used in the building of the pyramids, but this doesn't mean that we are unable to work out how it was done.

It should also be taken into account that although the pyramids were built to an incredible degree of accuracy, they are *not* perfect. The base of Khufu is a few inches lower on on one side compared to the other. The angle of slope is not a precise 52 degrees as some claim, it is a little under 52 degrees, as near as can be ascertained with only crude blocks left instead of the original smooth surface. The length of the four sloping edges all vary by a few feet. Very accurate - even amazingly accurate - yes; perfect, no.

In order to try and establish who built the pyramids we have to examine the evidence that we have. We have only the pyramids themselves, the excavation sites where the blocks were quarried from, and historical accounts.

The evidence:
Lets start with the excavation site. It looks pretty much like any other quarry you might see today, except there is obviously no machinery. At the quarry face there are blocks cut into the rock but not yet cut away. There are rough hewn blocks scattered around ready for transporting and on-site finishing. The entire quarry shows obvious signs of systematic development of cutting blocks out from the face and transporting them from site. The rough hewn free standing blocks show the scars of repeated chisel blows where they were chiselled out of the rock face. There is nothing in the manner of these blocks that is anything other than old fashioned quarry work using a mallet and chisel. Nowhere is there any sign of advanced technology having been employed, just the opposite. The blocks were hewn out of the rock-face by manual labour, the signs are unmistakable. The chisels used were made of copper, the hardest metal then available, but even they were only good for about 100 blows before blunting, even though limestone is relatively soft and easy to work compared to hard rock such as granite. As the chisels were blunted they were

exchanged for re-sharpened ones, and the process was repeated with a team of blacksmiths constantly re-sharpening and tempering the chisels.

How were the blocks transported to the pyramids? By man power. The vast majority of the blocks weighed in the region of 2.5 tons and were transported on wooden sledges. They could of course have chosen any size for the blocks, but this must have been the optimum size, any bigger would probably have slowed them down. A team of men with ropes could drag the sledge across the clay floor, and this could be eased with a little water tipped in front of the runners helping the sledge to slide easier. It is estimated that it may have taken 10 years just to build the ramp from the quarry to the pyramids. In this manner all the blocks could be transported to the site of the pyramid without presenting any insurmountable challenges. So far no alien technology required, it could all be done by well organised teams of men, and a great deal of manual labour.

The work force was was not one of slaves, the Egyptians didn't need slaves. The Nile supplied a very fertile land where farming was relatively easy and food abundant. This civilisation had time on its hands, no wonder they were such great mathematicians, astronomers and architects. The work force was primarily made up of farmers, recruited nationwide for a period ranging from a few months to a few years, and they served their time for their king, much like serving National Service today in the armed forces. A total of 20,000 men would have been needed for the task, ranging from unskilled hauliers, semi-skilled quarry men, skilled quarry men, masons who finished the blocks, men who placed the blocks, officials and caterers. Roughly, this amounts to 4000 actual builders, and a supporting cast of 16,000 ranging from mortar mixers to caterers. A village was purpose built to house them all and they were well fed and cared for in return for their work. The remains of the village can still be seen today.

Now for making the pyramids themselves. About 2,550 B.C., King Khufu, the second pharaoh of the fourth dynasty, commissioned the building of his tomb at Giza. Some Egyptologists believe it took 10 years just to build the ramp that leads from the Nile valley floor to the pyramid, and 20 years to construct the pyramid itself. Having man-hauled the blocks to the site of the pyramid the obvious problem now is how to stack them up. There are a number of ways this may have been achieved, all of which require a ramp, or a system of ramps, as the only method available to the ancient Egyptians was man-power, and they had that in abundance.

The actual method of ramps used is not known with any certainty, but it most likely started with a single ramp by which means the blocks could be hauled into position. The blocks were laid down in layers, each successive layer being a little smaller in area than the one below it to give the pyramid its shape. As the blocks are laid onto a level surface, the same height as the ramp, no lifting was required, only hauling of the sledges. Removing the blocks from the sledge may have involved no more than dragging the block off the sledge. The entire pyramid could have been built using this simple system without a single block actually having to be lifted off the ground! No alien technology required. Very smart people these Egyptians.

As each successive layer was laid the ramp would need to be heightened and lengthened so as not to be too steep. Eventually this method would reach a limit where the size and construction of the ramp would be nearly as complex as the pyramid. The easiest way around this problem is to curve the ramp around the pyramid as the pyramid increased in height.

Finally, all that remained was the placing of the top stone, followed by the placing and fitting of the smooth white blocks. As the facing stones were placed so the ramp could be removed as they worked their way back down. The pyramid required a certain amount of interior design and construction for the burial chambers, and this was no easy task. The blocks that protected the burial chamber were 50 ton blocks of granite. Even with the huge teams of men at their disposal and a system of ropes and overseers guiding them, it would have been a difficult and dangerous task. There are still marks visible on the blocks and in the interior of the pyramid that were used to guide the blocks into position. Difficult yes, impossible, no.

It was men that built the pyramids, make no mistake. Do not underestimate the intelligence of the ancient Egyptians or the trained manpower that was at their disposal. It was a colossal effort of team work taking 30 years to complete, the actual building of the pyramid itself taking 20 years.

How did they align the corners of the pyramid so accurately with the four points of the compass? This was the easy part. It just takes a curved wall facing more or less North as judged by the stars. Select a rising star on the Eastern horizon and mark a line on the top of the wall pointing to it. Mark another line on the wall when the same star is low on the Western horizon. Take the line straight down the wall, using a plumb line, and then extend it along the ground until it meets the other line. Bisect this angle and you now have a line pointing exactly due North.

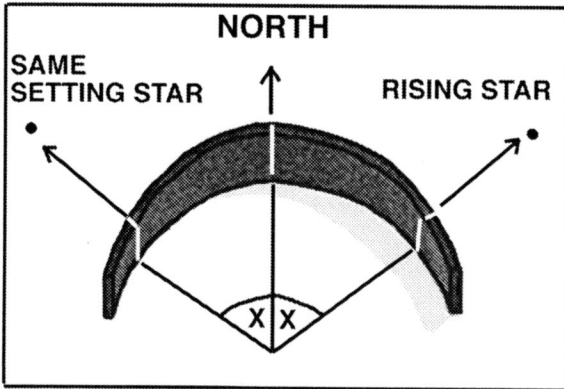

There are any number of very simple methods that can then be employed to construct a right angle that will then align with the other points of the compass. After that its just a question of placing markers in the ground for the four corners of the pyramid. As I said, that part is easy, it does not require alien technology, just a little brain power.

How did they achieve such a near perfect 52 degree angle of slope? Because they were clever! They were masters of angles, a skill they had acquired in part through astronomy, and also through a good knowledge of mathematics. Each mason that worked on finishing the outer blocks had a template with an angle of 52 degrees that he used to cut his block to fit. After that it was careful alignment of block to block as they placed them carefully on the inner blocks. It was literally done step by step.

Mystery solved, that's how it was done. Maybe not using that exact method, but something very like it. There is absolutely no need to suggest that it couldn't be done by man power, and certainly there is no need to go to the ridiculous lengths of suggesting it was done with the help of aliens. That idea is pure nonsense.

Keith Mayes

Why were the pyramids built? They were tombs for the king. The Egyptians had long observed a point in the sky around which all the stars appeared to rotate (the north celestial pole, today marked by the close proximity of the Pole Star) and believed the area to be an immortal place as these stars never set below the horizon and were always visible. These northern circumpolar stars were particularly reverenced as "knowing no destruction" and symbolised those who had triumphed over death and gone on to eternal life. The alignment of the pyramid to the compass points was to align it with the 'immortal place' in the sky, and the shape was believed to represent the rays of the Sun. The pyramid was in fact a device designed to transport the king to the immortal place so that he may live forever, as would all the workers that assisted in its construction.

N.B. The theory of how the pyramids were actually built is not my own pet theory, but one considered by experts in the field as to be the most likely. It is a known fact how the stones were carved out of the quarry. Graffiti inside the pyramid reveals that teams were used in competition with one another, with the best performers receiving bonuses of extra beer or days off. The workers village is a known fact. The ramp from the quarry to the pyramid is a known fact. That the huge work force was recruited is a known historical fact. The labour force required and the time it would have taken has been very carefully calculated. The type of ramps used is open to question, but all are agreed that ramps were used, there is no other way it could have been done. And it was done - by men.

21. Astrology: It's in the stars?

"Oh, what a tangled web we weave when first we practice to believe."
Laurence J. Peter

Here we are in the 21st century, in the age of education and enlightenment (?) where a large proportion of the planet's population will consult their horoscope before deciding on what course of action to take. To me this has as much relevance to reality as killing a chicken and spreading its entrails on the ground and taking a 'reading' from the resultant sticky mess, as indeed they do in some cultures.

I must confess that I am at a complete loss to understand this strange belief in astrology. Even stranger perhaps is that whenever I have questioned those who do believe in astrology, they have been unable to offer any explanation as to how the position of the stars at the time of their birth can influence the type of personality they will have, or how the stars will determine their 'lucky' numbers or 'lucky' colours. Just some vague notion that "it must have some effect on us". 'Vague' being the operative word.

So why do they believe it? They themselves do not know, and I'm sure I don't! Some will tell you that "it has something to do with gravity from the stars, where they are in the sky will affect you". The truth of the matter that the position of the midwife will have a far more powerful 'gravitational influence' than the stars goes unheeded. Similarly, if you were born at the foot of a mountain range that would have a 'gravitational effect' on you that would be absent if you were born in the flatlands. Why do they not consider that this 'gravitational attraction' would have a far greater effect by an enormous magnitude, then any from the distant stars could possibly have?

I believe the answer lies in the stars. I think people who believe in astrology in some odd way imbue them with a strange mystical power. People who have strange beliefs in 'mystical powers' can not be persuaded by logic alone that it is a falsehood, they believe it and that's that. It could perhaps be argued "So what, there's no harm in it". Isn't there? To allow a popular misconception to guide you through life, often when you are at your most vulnerable, harmless? You could be 'guided by the stars' into making the worst decision of your life, and when it all goes horribly wrong, going back for even more ill judged advice. If you need advice then seek it from your friends and family, not some astrologist who will peer over charts and diagrams and then pronounce what is in store for you. You have to make the best of what life offers you, and when you need to make a tough decision, think about it rationally and logically. Try to judge the outcomes each choice may bring. Seek expert advice, speak to others who have been in a similar situation, speak to friends, self help groups, professional advisers. Do not consult astrologists. They are selling snake oil.

I watched an interesting TV programme recently testing star sign personalities. 30 members of the audience were all given the same personality description, the sort that astrologers give you. The audience were then asked which of them believed the description fitted them exactly. To be honest, I can't remember the exact figure, but it was most of them, around three quarters. Why is that? It was because the description was so generalised, and contradictory, it would just about fit anyone. Typically they say "You enjoy company but at times wish to be alone", or "You are capable of making quick decisions but sometimes feel uncertain".

The main problem with astrology is in its claim that the position of the stars at the time of our birth somehow affect us. I have never been able to find anyone able to explain how. All the performance with charts and 'your Moon is in the house of Libra, with Jupiter in ascendancy" is meaningless unless supported by an explanation of how this affects any of us. If the positions of the stars and planets affect us, then how do they affect us? I have been told variously that it's to do with our planets momentum, the point where it crosses the Sun's orbit, whether or not we are moving away from or towards the galactic centre, and other weird explanations, but never, ever, any explanation as to *how*.

History
It is not too difficult to imagine how astrology came into existence, and a look at the night sky today does offer us a clue. The Crab Nebula, also known as Taurus A, M1 and NGC 1952, is a glowing cloud of gas and dust in the constellation Taurus and is the remnant of a supernova explosion that was observed by Chinese astronomers, and others, in AD 1054. At the time of the supernova it was temporarily brighter than Venus, being visible in daylight for 23 days. A supernova is the explosive death of a star in an event so violent that for a brief period that single star shines as brightly as a whole galaxy of more than 100 billion stars. The Crab Nebula today is a beautiful site when viewed from a dark sky location even through a very modest telescope.

This supernova was well documented by the Chinese astronomers for two main reasons. One reason being that they were keen observers of the night sky - keen astronomers - and astronomy has for thousands of years served the very practical purpose of being an excellent calendar. The ancient Egyptians, for example, used astronomy as a means of predicting the annual flooding of the Nile. When a particular star rose above the horizon at particular point that was an indicator that on average the Nile would flood within two weeks. This would be because the star always reached that position the same time each year of course.

It is not then so surprising that in using the position of the stars to predict the flooding of the Nile and the changing of the seasons, that the position of the stars also came to be used to predict future events that were not tied into the regular changing of the seasons.

Returning to the Crab Nebula and the Chinese astronomers of AD 1054, they were not only astronomers, but also served the dual role of astrologers. The astronomers/astrologers would look for any changes in the night sky - comets, supernova, asteroids - anything that was outside the normal regular predictable pattern. These unusual events were then interpreted by the astronomers (astrologers) in the light of current events and taken to be predictions. For example, an event such as the Crab Nebula supernova could be taken as a sudden change in events, such as the start of a long awaited attack by a rival army. The astronomers would then warn the emperor and thus confirm their position as powerful mystics with the ability to foretell the future. It is not difficult to imagine how by keeping an open ear and watchful eye, the astronomer/astrologers would be able to make accurate 'predictions' on a fairly regular basis, especially self-fulfilling prophecies. Perhaps this is how the belief in astrology grew, from simply using the position of the stars as a calendar to using them to predicting annual seasonal events - such as the flooding of the Nile - to predicting events that were not related to the seasons.

Can there be any possible scientific explanation for why the belief in astrology could be correct?

A scientific explanation?

Dr Percy Seymour, a British astronomer, claims the planets may exert an effect indirectly, by stirring up magnetic activity on the Sun. He says that during the 1960s, researchers at NASA found evidence that when the planets were in conjunction or opposition with the Sun - that is, in line with the Sun, looking either towards the Sun, or directly away from it - solar magnetic storms became particularly violent. Dr. Seymour argues that this may in turn create outbursts of fast-moving particles from the Sun, the solar wind, which are known to affect the Earth's magnetic field. It is these changes in the geomagnetic field that Dr. Seymour believes can affect humans at birth, affecting their nervous systems, and altering their personality.

All very interesting, but it does not tie in with the beliefs of astrologers. They claim that our personalities are very similar to others who are born at the same time of year, such as all those born during the period 20th January - 18th February, the sign of Aquarius, having similar personalities. This is relating our personality types to only one thing, the time of year we are born, which has no connection at all with any changes in geomagnetic fields, which are completely random and unpredictable. I am very interested in these changes, and keep a close watch on my magnetometer which measures these changes, as they may herald the start of the 'Northern Lights' - the aurora created when high energy particles ejected from the Sun interact with our atmosphere.

Personality types
   Could there be any explanation as to why the time of year could affect our personalities? Perhaps the alignment of the stars has nothing to do with it, other than indicate the time of our birth, and in reality this is what matters? This could mean, for example, that it isn't the stars that are having an effect on us, just the time of year. Perhaps it's hours of darkness, or the average temperature, the weather, the height of the Sun above the horizon? But again, these things are not consistent. A person could be born in Scotland say mid winter and the weather could be a mild and sunny with a temperature of 12 Centigrade, and stay that way for weeks. A person born in the same house exactly a year later may be born during a blizzard and temperatures of -20 Centigrade. Yet even with these vast differences in local conditions these two people are expected to have the same personality because they share the same birthday. Doesn't seem logical does it. If not the weather due to the seasons, then what else could it be. In the example given the only thing that was the same for both births was the position of the Sun, both as seen in the sky and in relation to the Earth's annual orbit around it. The Sun however, is just a star like any other, we just happen to be very close to it. The position of the Sun above the horizon cannot be the factor that influences our personality, it will be at very different altitudes above the horizon in the far north of Scotland compared to the far south of England.

We can rule out any connection with the position of the Sun in the sky just as we can with the time of the year and the seasons. What else is there?

When astrologists create an astrological chart for an individual, they use that person's date and place of birth to show the position of the stars and planets as they appeared in the sky at the time of birth. This is very simple to do. There are a number of very good computer programmes that can be purchased by keen astronomers, such as myself, that will reproduce the night sky from any location in the world, or even off it, for a period covering a few thousand years into the past to a few thousand years ahead. Very useful for discovering what is visible in the night sky from your location. So armed with the knowledge of how the sky appeared at the moment of your birth they then proceed to tell you how your life is mapped out. It is interesting to note that in order to do this they need to know where all the planets in our solar system are at the moment of birth in relation to various designated areas of the sky. The only conclusion that can be gained from this is that it is the position of the planets that is important, not of the stars. What effect could the planets have on us? The only possible effect is that of gravity, nothing else about the planets could have any effect upon us, they are just lumps of rock orbiting the Sun just as planet Earth is. The problem here is that the gravitational influence of the planets on us is so minute as to be undetectable, is far less than the gravitational effect of the position of the mid-wife standing next to the bed, and completely swamped by the gravitational effect of the Moon. That can cause the ground beneath your feet to rise as much as 40 cm when it passes overhead! We can safely rule out the position of the planets.

Could the position of the Moon have any astrological bearing? Some people will tell you that the Moon does affect our personalities, especially at full Moon, hence the term 'lunatic'. However, statistical crime records show that there is no change in the behaviour of people at the time of the full Moon, despite it being a commonly held belief.

Nurses and doctors who work in hospitals caring for the mentally ill do not report any difference in the behaviour of their patients at the time of the full Moon. It is just one of those old urban myths. What force could the Moon exert on us anyway? The Moon, like the planets, is just a lump of rock. It is in orbit around the Earth and exerts a strong enough gravitational force to raise and lower sea levels and the continents. At the Bay of Fundy, Nova Scotia, for example, the tidal range is a massive 40 feet, and the continent of Europe is lifted 40cm (16 inches) each day as the continent comes directly face to face with the Moon. Other than that it only reflects sun-light onto us.

The Moon does have a gravitational effect upon us, although what changes that could produce in our behaviour, and our future actions, is unknown, if any! It is difficult to imagine how the gravitational force exerted upon us by the Moon at the time of our birth could have any influence on our personalities or future actions.

We have examined a number of possible explanations as to why the time of our birth may have some influence on us, but have been unable to discover any consistent or logical explanation. We have ruled out the position of the Sun, Moon and planets as having no bearing on the matter, at least that we can detect. We have also ruled out any connection with the time of the year as having any connection with the type of person we are. What else is there? There is nothing else that I am able to think of. It may be argued that there is a force that has a direct effect upon us, but the nature of this force has so far eluded us. It is not possible to argue against this hypothesis, but that is not the point, the point is where is the evidence to support such a wild claim?

To support such a claim would require evidence that the world's population can be roughly divided into twelve distinct personality types depending on the time of year they were born, and there is no such evidence. Even if such evidence were to be produced, that in itself cannot be offered as definite proof that the astrologers have got it right, it would only mean that the world's population can be categorised into twelve different personality types.

Predicting the future

Apart from defining personality types determined by a person's date of birth, astrologers claim to be able to predict events that will occur at some future point in that person's life. This should be very simple to test. To begin, I went to a web site, gave all the relevant information concerning my date and place of birth, and was given the following forecast:-

"Jupiter conjunction Pluto: Self-betterment

1 March 2003 until 2 March 2003: This influence brings the urge to achieve to the forefront of your life. You will make great, even extraordinary, efforts to gain success as you personally define it. You work harder and strive to gain your objective with every ounce of energy at your disposal. Consequently, this influence often occurs just at the moment when some tremendous effort in your life bears fruit.

Sometimes there is a tremendous drive to gain power, and this influence can bring you power even when you are not trying especially hard to get it. Remember that each person has opportunities for a different kind of power. It may be great or relatively humble in terms of the larger society, but it will be meaningful to you."

As it happens, this 'prediction' was already two weeks in the past, and thus allowed me to test its accuracy. I have to say that it bears no relationship at all to how I felt during that period, about 'great efforts to gain success'. But notice how later on it employs the usual trick of contradiction. 'It may be great or relatively humble in terms of the larger society, but it will be meaningful to you.' In other words, I may, or may not, make great efforts to gain success. Brilliant!

In order to test the accuracy of a prediction it is of course necessary for the prediction to be clear and unambiguous. Having taken the trouble to check many 'predictions' I soon came to the unsurprising conclusion that they do not meet these requirements. The example of mine that I have given is typical example of so called predictions. They tend on the whole to be based mainly on emotions and feelings, which are of course very subjective.

I would be a lot more impressed if instead of vague and contradictory 'predictions' about one's emotions and state of mind, a definite prediction of some event were made. If for example an astrologist were to predict that next week I would trip over and break my arm, I would be very impressed if that event actually came about. However, astrologists do not make that sort of definite and testable 'prediction'. I wonder why that is?

Free will

If we were, for the sake of argument, prepared to accept that it is possible to predict the future, then we would have to accept that we do not have free will. We would have to accept that our lives are pre-ordained, that what will be will be. That being the case we may as well all stay in bed.

In conclusion it has to be said that astrology has no evidence to support it whatsoever, not statistically or experimentally, nor even have a working hypothesis. Those that believe in it do so simply because they wish to.

22. Telekinesis. Fact or fantasy?

"I never cease being dumbfounded by the unbelievable things people believe."
Leo Rosten

Does the human mind posses the ability to move objects? Silly question really, of course it does, it moves our bodies. But can it move objects other than our own bodies, by the power of telekinesis? I know for a fact that I do not posses this ability but others do make the claim. What evidence do we have? In order to answer this question it must first be decided what we will be prepared to accept as evidence of telekinesis.

We can rule out photographic and film/video evidence straight away, for obvious reasons. Eye witness accounts? No, simply because no matter how sincere the eye witnesses, they could have been deceived. This then only leaves one sure way we can be certain that telekinesis has taken place, it must be performed in front of us, live. Not on a stage in front of an audience of hundreds, that's no good either. The world famous illusionist David Copperfield under those conditions can levitate a cage of lions! It has to be carried out in front of a few people close up and under strictly controlled conditions. The room to be used must be open to inspection before and after the experiment. All equipment in the room and the object to be moved by telekinesis must be open to examination before and after the experiment. All witnesses must be allowed to use any passive testing equipment they want during the experiment, such as infrared cameras, high speed filming, magnetometers, etc. The experiment must be set up in such a manner that the object to be moved must not be able to move by any other means, draught, gravity, an unstable arrangement, etc. and preferably the direction to be moved must be clearly stated in advance.

Under those conditions, how many examples of telekinesis do you suppose have been recorded? Go on,

have a guess. The correct answer is NONE. I went to Google and typed in "telekinesis' and a choice of sites was offered up. I don't recommend you bother but go ahead if you really feel the need. I dutifully worked my way through the first half dozen or so and then gave up in despair, they were absolutely awful. Most were just full of the usual predictable mystical mumbo jumbo. Some claimed to describe tests that were performed in 'scientific' labs, but when you actually cut through all the hyperbole and quotes, you are left with nothing of substance. Most of them referred to other experiments that supported their findings, but again, hard facts were thin on the ground. To be accurate, there was not one single 'experiment' that used any halfway decent scientific method of reporting the experiment, and from the scant information given (although much dressed up) it would be impossible to repeat the test because not enough is known about how it was actually set up. In other words, some people may find these sites to be of passing interest, but none of them will add one jot to their knowledge. Okay, perhaps I should have persevered and checked out some more sites, after all, a sample of just six is hardly conclusive. No way! I prefer to keep my brain uncontaminated, you go ahead if you want, and if you find anything useful perhaps you could let me know. Good luck. (Since first writing this article I have checked out a great many more sites on the subject and have still not found anything of even the remotest interest.)

For those of you that claim you posses TK, here is some very exciting news for you! The James Randi Education Foundation is offering a million dollars to anyone who can prove paranormal, supernatural or occult powers under test conditions. The money is still waiting to be claimed as at August 2003. Here is the web site for you so that you can claim the money: http://www.randi.org/research

I receive emails every now and again from people claiming that they have TK. My reply is always the same. First claim the million dollar prize and then contact me again! These same people offer up a variety of reasons why they have not claimed the one million dollar prize. Such as: 1) The money is not worth the publicity it would bring. 2) They know it is a waste of time trying to prove it to those who refuse to believe. 3) They do not feel it is necessary to prove their abilities, they know they can do it and that's enough. 4) They were not blessed with this wonderful gift in order to abuse it by making money. (Now where have I heard that one before?) Okay, we get the picture.

In case you are interested in knowing the sort of thing the average TK web sites offer, here is a typical example, in all its glory:

http://members.aol.com/dstnysangei/pw.html (As at July 2003)

The following is an exact copy of the page.

Psi Wheel

This is a simple experiment to exercise telekinesis.

Take a metal thumbtack (the kind with a flat, circular base) and set it so that the sharp end is pointing upward. Now take a small square of white paper, about 1/4 inch by 1/4 inch (maybe even smaller), and carefully balance it on top of the tack. You can replace the bit of paper with a piece of carefully cut aluminum foil if you feel more comfortable with metallic energies.

Meditate for half an hour.

Now sit in front of the "psi-wheel" you have just created. Stare on one corner of the square of paper. Focus all your energy on that one corner, willing the paper to start turning on top of the point. Imagine the wheel (the paper) begins to spin on the axis (the tack). Hold your concentration for as long as it takes for you to get it spinning.

Once you have the hang of this, try knocking the paper off the tack!

See what I mean? Put a very, very, very tiny square of paper onto a sharp metal point and wait for it to move. When it does you have moved it by the power of TK! My

God! People actually take notice of this rubbish? Really? Why?

So this experiment is set up in such a way that any movement of the little piece of paper (sorry, Psi wheel) cannot possibly be due to air currents? Or vibrations? Or your breath? or any other external factors at all, other than your mental energy? Oh! they could, but nonetheless it is definitely due to TK for sure. I see. Well that's cleared up that one then!

In order for this 'Psi wheel' experiment, and others like it, to have any meaning at all, they need to be conducted under strict test condition.   Let's use the Psi wheel experiment as an example. Firstly it would be necessary to make sure that all other forces that could move the piece of paper are, as far as possible, prevented. We could perhaps place the Psi wheel inside a screw top glass jar on a thick bed of sand. The sand will help damp down any vibrations and once the lid is screwed onto the jar all outside air currents will be eliminated. The jar should be placed in a position where no vibrations are normally detected, and if conducted at home, then a time should be chosen when no other people are in the house.

A control will then need to be set up. The control is a duplicate of the test equipment so that results can be compared. The control should be set up in another room away from the test equipment. An observer will be required to observe each of the Psi wheels. In this particular experiment however, as the observer may exert an influence on the control Psi wheel, it would be better if it were not directly observed but videoed for later analysis.

The experiment should then begin with the subject attempting to move the test Psi wheel. After a given time a note should be made of any movement of the Psi wheel, if any, and this compared to the movement of the control. Ideally, in order to 'prove' TK had taken place, only the test Psi wheel should have moved. If, on the other hand, they both moved pretty much the same, then there is no case for the subject to claim that it was TK making the test Psi wheel move, for if so what moved the control Psi wheel? However, this is just the beginning, we have only recorded the result of a single test, hardly conclusive evidence either way.

The test will now need to be repeated at least 100 times and results compared with the control. The next step is to test the results using a variety of statistical methods in order to establish if the number of times the test Psi wheel moved, if at all, is statistically significant. If having done all this a level of significance is achieved that suggests that the results are above what may be expected by pure chance alone, then these findings, together with precise details of how the tests were conducted, can be given to another group for verification. This second group will duplicate exactly your experiment and compare results. If your results were meaningful then their results should be very similar to yours. The stage has now been reached where the first group may have sufficient confidence to publish their findings and risk their professional integrity, not to mention their careers. Other groups will now carry out their own experiments and all the results collated. Ideally they would all be very similar in their findings, and so confidence will be steadily increasing. Eventually, after many similar experiments over a long period of time, the results will be treated as being correct. This, by and large, is how experiments are carried out.

The next time someone feels they need to contact me because they have moved a Psi wheel by TK, I urge that they read the above paragraph again and think again!

One of the things I have attempted to discover is the nature of the this force that is used to move objects. I have found no explanation for it as yet. Some suggest that it helps to put your finger very, very, very close to the tiny piece of paper (sorry, Psi-wheel) I bet it helps! This would suggest that the force emanates from the fingers. Others suggest that you only concentrate and allow the force to do its thing, suggesting in their writing that it emanates from the head. I must confess to finding the nature of this force very puzzling indeed. What manner of force, or energy if you prefer, is it that can penetrate through the brain, skull, nerves and skin tissue without causing any damage or pain or disturbance whatsoever, be capable of travelling through the air, again without having any effect upon it, yet when it reaches the target object 'knows' it has reached the selected object and then, and only then, have any effect? Amazing isn't it! We have a very selective force here that passes through some substances but not others, and then packs a punch!

I did come across one 'definition' of what TK is and it went like this: *"Physical energy is created by electromagnetic impulses. Universal life force energy, or psychic energy, is called Chi. [Chee] In telekinesis one taps into Chi energy then combines it with physical energy."* So that you have it, a complete description of what TK is. Impressed? No? How about this then. I found a web site that gave these two experiments for improving your TK abilities. The first involved a compass and concentrating the mind to make the compass needle move. The second involved suspending a needle on a thread inside a glass jar and again concentrating the mind to make the needle move. How many of you can spot the obvious connection? Yes, of course, both needles are affected by the Earth's magnetic field. Both will also be affected by the movement of nearby metal objects. The needle in the jar experiment interested me because it is so similar to my home made magnetometer that I use to detect magnetic storms, known as aurora, or Northern Lights. My magnetometer is nothing more than a small bar magnet glued to the base of a small square mirror that is suspended from a thread inside a sealed glass jar, to prevent air currents moving the mirror. Magnetic storms cause a disturbance in the local magnetic field which deflects the bar magnet a little from its normal magnetic north/south alignment, and by shining a laser pointer light onto the mirror and noting the reflected position of the spot on the wall I am able to tell when a storm is building or in progress by the amount of the deflection. However, it is first necessary that I take into account the normal daily cycle of movement that has nothing to do with any unusual magnetic disturbance but simply the regular pattern of daily change caused by the rotation of the Earth and the elliptical shape of the magnetic field over the north pole.   Did you notice the magic words there?   *"It is first necessary that I take into account the normal daily cycle of movement that has nothing to do with any unusual magnetic disturbance..."*   This is a very important factor that TK supporters studiously ignore.   **The needle will move anyway regardless of whether or not you are sitting there for hours on end trying to make it move!** Oh dear oh dear! You have to laugh don't you!

We have all seen magicians and illusionists perform amazing feats, and although we know it to be a trick we are nonetheless at a loss to understand how they do it. How on earth does David Copperfield make a train disappear into thin air, a huge great train with carriages! I bet there are hundreds of magicians out there who could move an object 'by telekinesis' in front of a group of scientists, and convince them. It just goes to show how difficult it is to prevent fraud, and how easy it is to convince eye witnesses that they have seen the real thing.

Have I ever seen anyone move anything by telekinesis? No. Have you? Are you sure?

I am absolutely certain that telekinesis is nothing more than a myth. Why? Because if people really could do it, and prove it, the question of its existence would not arise, we would all know it to be a fact. The fact that it is a very questionable subject should suggest to everyone that it has a problem.

One of the sites I checked out related the case of a very famous (to a small handful of people) exponent of the art who could move assorted objects, such as a salt cellar for example (?) by thought alone. However the article went on to say that when there were sceptics in the room the experiment may not work! Nice one. Since when did the laws of physics require that we have to believe in them in order for them to work? These things are self evident, such as the fact that a rock will fall to the ground under the influence of gravity. It does not require that the rock believes in gravity.

If a thing requires that you have faith in it in order for it to work, then it is no more than a self manufactured unsupportable belief. Facts do not require faith.

One example of a TK 'experiment' I found on the Net described the case of a frog's still beating heart in a jar of solution (What psycho thought that one up?) that the gifted person could stop and start beating at will. I mean, come on, it couldn't possibly just stop and start anyway? After all, it is rapidly dying! This stuff is just unbelievable. Another described a person moving the flow of water jetting up from a tiny fountain. Really impressive or what! Not that it would move about anyway. Now you know why I gave up. I even sent an email to a guy who claimed he could move stuff about, and on his site showed a photo of himself in his 'lab' with his 'measuring equipment'. He gave a vague description of what he was doing. I asked for some more details. Guess what? No reply.

If you believe you have the power to move objects by telekinesis, then that's just wonderful for you, I am so very pleased for you, but please do not bother to contact me, as some already have, until you have claimed the million dollar prize, and followed the rigid rules for conducting experiments as previously described.

To those that contact me every once in a while to claim they can move objects by telekinesis, but not when observed by those who disbelieve, and so are unable to claim the million dollar prize, I know just how you feel. I can walk on water when no one's watching, and get 20 strikes in a row ten pin bowling. No one believes me either! Strange that, I wonder why?

Come on people, telekinesis is pure nonsense! If you think you can do it then you are only deceiving yourself, the rest of us just don't buy it.

23. Crop circles. Who makes them?

"Just as you cannot do very much carpentry with your bare hands, there is not much thinking you can do with a bare brain."
Bo Dahlbom *"Computer Future"*

That crop circles exist is a known fact, the question remains however as to who or what makes them.

Crop circles were not known prior to 1970, apart from the one reported exception of the *Tully, Australia Saucer Nest of 1966.* Crop circles first appeared in the UK in the 1970's, starting with simple circular patterns and developing over time into huge and complex geometric formations. In 1991 two elderly landscape painters named Doug Bower and Dave Chorley confessed that they had been making crop circles in English grain fields since the 1970's after reading about the Tully, Australia Saucer Nest of 1966. The pair demonstrated how they did it for a film crew and told how they had devised the idea over a pint or two at their local pub. It would appear that the Tully, Australia Saucer Nest of 1966 is the earliest reported crop circle, although it is not what may be recognised today as a typical crop circle.

What is the Tully Saucer Nest story?
The following article is from http://ufos.about.com (August 2003). I have reprinted the story as it appears in all its glorious entirety so that you make up your own mind and not accuse me of putting my own slant on it. This is it, where it all started:

"At 9:00 am on January 19, 1966, a calm sunny day, a 28 year old banana farmer named George Pedley was driving a tractor near Horseshoe Lagoon on the property of Albert Pennisi, near Tully, in tropical far north Queensland, Australia. When he was about 25 yards from the lagoon, he heard a loud hissing sound above the noise of the tractor.

Suddenly, an object rose out of the swamp. When I glanced at it, it was already 30 feet above the ground, and at about tree-top level. It was a large, grey, saucer-shaped object, convex on the top and bottom and measured some 25 feet across and 9 feet high. While I watched, it rose another 30 feet, spinning very fast, then it made a shallow dive and took off with tremendous speed. Climbing at an angle of 45 degrees it disappeared within seconds in a south-westerly direction…

Another surprise came when Pedley rounded the bend of the road and came to the spot from which the object had risen. There in the lagoon was a large circular area that was clear of reeds and in which the water was rotating slowly. It had not been like that three hours earlier when he had passed the lagoon. After looking around, he got back on the tractor and left.

A few hours later, at about noon, Pedley returned to the lagoon for a second look. The scene had changed, because now the circular area was covered by a floating mass of green reeds that were distributed in a clockwise radial pattern. The circular mass of reeds was about 30 feet in diameter.

Pedley was by now excited enough about what he was seeing to go and tell Albert Pennisi, the owner of the sugar cane farm land on which the lagoon was located, and another friend. Pennesi recalled that his dog had acted strangely that morning, barking madly and heading off toward the lagoon at about 5:30 am. Pennisi and the other man were amazed by the circular mass of reeds. Wading out to the mass, they found that they could swim under the mass of reeds and that the lagoon floor beneath it was smooth and showed no traces of roots. Oddly, the outside edges of the mass of reeds angled down, similar to the shape of a saucer placed face down. Pennisi went and got his camera and took photographs of the mass of reeds, which was now beginning to turn brown on its top surface. George Pedley reported his experience to the Tully police that evening, and they in turn reported it to the RAAF after making a trip to the site the next day, January 20.

Within days, the media had picked up the event and the area was filled with investigators, many of whom were trying to prove theories as to the cause of the "nest" such as helicopters, big birds, crocodiles, reed-eating grubs, and whirlwinds of one sort or another. Pedley's UFO sighting was all but overlooked in the flurry of explanations. During the course of the investigations, as many as five other "nests", all smaller than the original, were discovered. In some of these, the reeds were rotated in a counter-clockwise direction and a couple of them showed signs of burning in the center of the nest. Samples of the original nest were sent to Brisbane for analysis, but nothing unusual was detected. Other than being part of the "nest", the only unusual thing about the reeds was that they turned brown in about 8 hours, whereas reeds uprooted by hand in the lagoon took three days to turn brown.

In another unusual twist, Albert Pennisi told a reporter from the Sydney, Australia newspaper The Sun that he had been dreaming about a UFO landing on his property for a week:

I'd get them almost every night. And they were beginning to worry me. I couldn't understand them. It was always the same. This thing like a giant dish would come out of nowhere and land nearby. And I would watch it in my dream and get real afraid before it went away. Then on Wednesday morning about 5 o'clock my dog suddenly seemed to go out of its mind. It was howling like a mad thing and raced off towards the lagoon.

What happened at Horseshoe Lagoon? There was never any evidence that there were any helicopters in the area nor any demonstrated reason for one to be over the lagoon. There was no evidence that crocodiles made the nest and analysis of the reeds from the nest showed no trace of "reed-eating grubs." There was no known bird that would or could make such a nest in three hours.

The best explanation that the RAAF could offer was that the nest was created by a willy willy, a type of small whirlwind known to occur in the area.

Although a conclusive determination could not be made, the most probable explanation was that the sighting was of a "willy willy" or circular wind phenomenon which flattened the reeds and sucked up debris to a height of about 30 feet, thus forming what appeared to be a "flying saucer", before moving off and dissipating. Hissing noises are known to be associated with "willy willies" and the theory is also substantiated by the clockwise configuration of the depression.

However, such whirlwinds, except when they occur in the desert as dust devils, normally accompany thunderstorms, and although the Tully event occurred during the rainy season, January 19 was a sunny day with little or no wind. Pedley described what he saw as a blue-grey object shaped like two saucers face to face. This description doesn't sound like a whirling mass of swamp debris, and there was no fallen debris in the area where the dissipation would have occurred. Finally, how does the whirlwind explanation account for the fact that the water was clear when Pedley looked the first time, yet was covered by the mass of reeds when he looked again three hours later?"

So there we have it, the first reported 'crop circle' story, all about a flying saucer making a reed bed to nest in! Either that or it was made by a willy willy! And what an industry that very silly story has launched, especially the tourist industry. There are are now companies that fly in tourists, many of them Americans, to the English countryside 'hot spots' and give guided tours of the crop circles. I wonder if these tourists are aware that there are crop circle groups, such as 'Circle Makers' that specialise in making crop circles? They will even make them to order! 'Circle Makers' have produced such 'circles' as an advert for Weetabix and a portrait of Richard & Judy, popular TV presenters in the UK. Circle Makers have their own web site at http://www.circlemakers.org. In case you may wish to order one.

Impressive though these man made crop circles are, it is not definitive proof that *all* crop circles are man made, you will have to make up your own mind on that score.

In my determination to find a good explanation for crop circles - other than the man-made ones - I trawled the Net looking for information, and soon came to the conclusion I wished I hadn't bothered. Theories abound. There are theories that cover Whirlwind Vortex, Plasma Vortex, Earth Energies, Extra Terrestrial Origin, Underground Archaeological, Hoaxes, God Force, Military Experimentation, and so on.

Apart from Hoaxes, they all seem ludicrous to say the least. If you wish to pursue the matter further then go to Yahoo, for example, and type in "crop circles".

We can broadly categorise these theories into three main types: Natural phenomena, extra-terrestrial and hoaxes.

With regard to natural phenomena, although these theories contain lots of pseudo-scientific mumbo jumbo, they all fail to explain how natural forces could make such perfect geometric shapes in such unnatural patterns.

Extra-terrestrial? I don't think so! See my page "UFO's: Fact or Fiction? "Why would they do it? I suppose you could argue that after employing fabulous technology to travel through interstellar space, that upon arrival in Earth orbit they ponder on the best way to communicate with us and come up with the superb idea that flattening some corns of wheat to make pretty pictures in a field would be the ideal method. Great. Makes a lot of sense to us dumb Earthlings.

Hoaxes? Seems a pretty good bet to me. Yes I know that some of the patterns look a bit tricky, but there are some pretty smart people about. To say that these patterns are beyond the ability of mankind is a bit of an insult to the intelligence of mankind. Why they would want to do it is another matter. Anyway, who is to say the photographs are not faked? Some that I have seen look decidedly suspect to me. I even have a popular crop circle photo on my web site and have received an email to say that it was shown on a TV documentary with two guys showing how they faked that photograph.

I find it odd, to say the least, that people refuse to accept that crop circles could be man made because of their complexity but are happy to accept that it could be the work of aliens. Circle Makers supply some useful 'insider' information on their web site including how long it took to make some of their crop circles. One formation made for The Daily Mail was 300 feet across and made in 4.5 hours. Another was 200 feet across and made by three people in 2.5 hours, and yet another was 218 feet across made in just 2 hours by only two people.

It is also interesting to note that Circle Makers say that it is easy to see that crop circles are man made because of prominent construction lines that underlie the paths of flattened crop that circlemakers use to lay out the geometry of the design.

Taking the opposing view that the crop circles are not man made we have the web site http://cropcircleconnector.com. This site tends to lean to the mystical and spiritual aspect of crop circles, stressing how in the main they appear around ancient sites such as Stonehenge, Avebury and Silbury Hill, all in southern England. They say that "Crop circle researchers are increasingly being asked to use creative and innovative thought as part of their investigation process. Some current field research uses meditation. lights and musical sounds in an effort to communicate new patterns into the fields as the circles phenomenon has been seen to interact with the human mind.".

A very different approach to Circle Makers!

So, to sum up, what do we have?

1) We know crop circles first appeared in the UK following that silly Australia Saucer Nest story of 1966.

2) We know that crop circles were started in the UK by two pranksters, basing their idea on the Australia Saucer Nest story of 1966.

3) We know crop circle started with simple circles and gradually, with practice, progressed to more complex geometric shapes.

4) We know we have Circle Makers who regularly make crop circles.

5) We know it is a simple matter to make large, very complex circles in a few hours under cover of darkness.

6) We know that man made construction lines can be found under the circles.

7) We know that some of the photographs are fake.

What else do you need to know?

CHAPTER 4

CONCLUSION

24. And the answer to Life, the Universe and Everything is?

"The effort to understand the universe is one of the very few things that lifts human life a little above the level of farce, and gives it some of the grace of tragedy."
Steven Weinberg

In a message from the creator?

The answer to 'Life, the Universe and Everything' is of course 42, as calculated by the super-computer in 'Hitch Hiker's guide to the Galaxy' by Douglas Adams. In the real world however, we look deep into space, peer down microscopes, smash atoms and compute complex equations, but it all amounts to the same thing - we are looking for answers. We are looking for a lump of rock, or a planet, or a single cell, that when examined reveals a message running all the way through it, and the message will say "Made by God", or words to that effect. If that were to happen we would know for certain that there was a God, that He created the universe, and that quite possibly we are here for a reason. Perhaps there really is such a message somewhere in the universe. Perhaps it's millions of light years away to ensure that we will not discover it until we have attained a level of technical sophistication sufficient to be able to understand it. Perhaps, on the other hand, it's right here, under our noses.

Where should we look for such a message should it actually exist? In the arrangement of the grouping of galaxies? In a grain of sand? In the bible? The problem is that if we look hard enough for a pattern, we can always find one because the human brain has a remarkable ability to detect patterns (I wonder why that is?). For example, let's take the bible. There is a book where the author claims to have discovered a secret code embedded in the bible text that reveals prophecies, and the book lists many of those prophecies that have come to pass, including the terrorists

273

attack on the World Trade Centre. The claim looks interesting at first glance, the author gives 'verified' mathematical proofs that the 'messages' are way above what could be expected by pure chance alone, and quotes endorsements from various mathematicians, etc. etc. Now here is the rub. The 'messages' are found in a pattern on the printed page, either up, down, diagonal, every second word, every third word, every second line, every fourth line - whatever, so long as the words are in close proximity. The bible text used is in Hebrew and the words are given a number of interpretations to meet today's modern language. A computer is then set the task of finding a specific event looking for key words, such as "War, Iraq, USA, 2003". Sure enough, given the broad scope, it will eventually find these words on the same page in some sort of pattern or other. However, it is not a prophecy when events can be found AFTER the event by searching for the key words. It is claimed that future events can be found, but with so much information generated by the computer it would be very difficult to isolate any meaningful event. I would argue that if, for example, this book 'Science, the Universe and God' was used instead of the bible, with the same amount of freedom of interpretation, I am sure it would produce the same results. Surprisingly, the book has sold well and a second book has just come onto the market with more of the same rubbish. It just goes to show how deeply people want to believe. It is even possible to purchase on the internet your own copy of the programme that searches for the bible code. However, simply searching for a message is clearly not enough, we need be very clear on what we are prepared to accept *is* a message.

We can look for 'messages' in the living world, and in a sense have already found the message of life in DNA. The information contained within that remarkable double helix is the blueprint for the formation of a living being.

Yet another fascinating example of scientists unravelling a hidden code is the human genome project. But both DNA and the human genome, although revealing much about life, as yet have revealed nothing else, and furthermore are unlikely to.

Could there be a message for us somewhere in the universe? I very much doubt it. If the universe had a creator, and wanted us to know, it would have been a simple matter to say, put the words "Made by God" on the surface of the Moon visible to the naked eye from Earth. Or if the creator wished us to be a little more educated before revealing his existence, he could have left a very teasing message in huge letters on the surface of the Moon -

3.141592653589793238462643383279502884197169399937510

which is pi to fifty decimal places. Now that would have got people really going! But the creator, if there is one, may have been much more devious in hiding the message, if such a message does indeed exist.

Instead of looking for a message in the bible, or in biology, or on the surface of the Moon, perhaps a message awaits us in the one science that is generally regarded as an eternal truth - mathematics. Carl Sagan exploited this idea brilliantly in his science fiction novel 'Contact'. Here the message was ingeniously embedded in the calculation of pi after a few million decimal places, the clear implication being that only the creator of the universe could have engineered this. In case you are wondering, pi has already been calculated to many millions of decimal places without showing any signs of a message, sorry. There is of course an almost endless list of possibilities where super fast super powerful computers may search for order in a strings of digits where no such order should be found. But what if they do find such order? What would it mean?

Suppose a computer in calculating an 'endless' number - such as twenty two divided by seven - after a billion decimal places 'found' the sequence:-

12233344445555566666677777778888888899999999910101010101010101010.

Would this signal a message, or a 'flag' that a message followed? First of all it would be necessary to determine what the statistical probability would be of producing that string by chance alone after a billion decimal places. There would then be endless arguments between various experts in the fields of mathematics and statistics as to whether or not the string was significant. This would no doubt be followed by every nut with a computer examining the sequence of digits before the string and after it and no doubt coming up with everything from the periodic table to the number 66 bus route time table. My point being that a message from the creator would have to be unambiguous, undeniable and unarguable. I just don't see a statistically unlikely string of digits tucked away in some obscure corner of mathematics as coming anywhere near meeting that criteria. So what would?

With our present level of technology I think it reasonable to say that no message exists on Earth that meets the criteria we have set. If it did we would have already found it, and if we discover something tomorrow that someone claims is a message then it will not have met the criteria.

Could there be a message out in space just waiting to be discovered? Just as an example let's imagine that the Hubble Space Telescope takes a very long exposure of an area of space that to the naked eye appears to be devoid of any object. After many hours of exposure the resultant picture reveals a group of distant galaxies, so far away that each appears to be a pin prick of light against the darkness of space. Nothing new in this by the way, only this time the galaxies are arranged in such a manner as to look like letters, and the letters spell out the message "Greetings to the people of planet Earth, the creator of the Universe welcomes you". I think that would stir things up a bit! Would anybody deny that the message was in fact what it purported to be? I am sure some would. They would claim that the image was fake, or that it was just pure chance the galaxies lined up that way when viewed from Earth, or that it was a computer programme glitch onboard Hubble, and so on. Those that choose not to believe are capable of finding all sorts of reasons why, just take a look at the Moon Hoax myth as an example.

So what message would meet the criteria of being 'unambiguous, undeniable and unarguable'? I think none, I think it impossible. We all have our own way of looking at things, our own way of working things out, our own standards. What to one may meet all the criteria, to another may meet none, it all depends on your personal point of view. So forget the idea of finding some sort of hidden message, it just wouldn't work! Unless that is, the message carried its own proof. If the message found by the Hubble Space Telescope for instance, was followed by the precise time of the next star to go supernova, and named the star, or the next gamma ray burst, or any other astronomical event that is impossible to predict, then I suggest that would have to be taken as absolute proof that the message was genuine.

If such a message were to be found, and if the message was undeniably genuine and proclaimed to come from the creator of the universe, what would we do? I would guess a lot more people would start attending church on a regular basis.

Suppose on the other hand we never find such a message, would we find our own answer to the meaning of Life, the Universe and Everything?

In a theory?

Perhaps instead of finding a message from the creator we may one day discover a theory that explains everything scientifically without any need for a creator. We may one day be able to explain all the forces of nature - even to the extent of explaining how the universe came into creation from nothing. If that ever came about, that the creation of the universe could be explained without the need of a creator, that even that the existence of a creator would be absolutely impossible, then that would be the end of religion. Or would it? Personally I doubt that very much. After centuries of belief I do not expect that the millions of people throughout the world who believe in a God of one sort or another will stop believing simply because some smart scientists come up with some clever maths! No, religion will not die just because of a theory, no matter how clever that theory may be.

Are we any closer to finding the fabled 'Theory of Everything'? This theory, TOE, carries the general name of 'Grand Unified Theory', or GUTS for short. The theory hopes to describe the physical behaviour of all particles and forces (the fundamental interactions) in one set of mathematical equations. There has been considerable progress made over the years in unifying the different forces. In the 19th century James Maxwell, for example, showed that electricity and magnetism were not two separate forces as first believed, but that they are two facets of the same interaction, now known as electromagnetism, and described by one set of equations. In the 20th century this description had been improved to include the effects of quantum mechanics. It is the hope and expectation of many physicists that all the four fundamental forces, in order of strength starting with the weakest - gravity, the weak nuclear force, the electromagnetic force, and the strong force - will be explained in one mathematical package.

GUTS would also require the combining of our two greatest theories. Ever since Einstein produced his remarkable theory of relativity scientists have searched for a single theory that would combine relativity and quantum theory. Quantum theory and relativity are the two greatest theories ever discovered, they are the great pillars of 20th century physics, each so accurate in its description and predictions that it cannot be faulted. But the amazing problem these two wonderful, precise theories have is that they are mutually incompatible, they cannot both be true!

Stephen Hawking in his 1988 best seller "A Brief History of Time" believed that the Theory of Everything would soon be found and that it would be the ultimate triumph of rational thought, saying: *"For then we would know the mind of God."* Hawking has attempted to unify general relativity and quantum theory, but in a lecture delivered in California and reported in the New Scientist dated April 3rd 2003 said:

"Up till now, most people have implicitly assumed that there is an ultimate theory that we will eventually discover. Indeed, I myself have suggested we might find it quite soon. Maybe it is not possible to formulate the theory of the universe in a finite number of statements. We and our models are both part of the universe we are describing. We are not angels who view the universe from outside."

Hawking is now re-examining the work of Kurt Godel, the Austrian-born mathematician. In 1931 Godel came to the remarkable conclusion that mathematics would never be 'finished' because there were theories that could not be proved from first principles. Mathematics was incomplete, he said. Hawking believes that he has come to the same conclusion for the physical world.

Some cosmologists have gone further than Hawking and said that *neither* relativity or quantum theory is true. Instead they can both be viewed as approximations of some bigger, more fundamental theory - the Theory of Everything. However, even if there is no such thing as a Theory of Everything, it may not matter much beyond the esoteric world of theoretical physics. Professor Richard Kenway of Edinburgh University, Scotland, said:

"Theoretical physics has always been looking for an ultimate answer to everything. From a philosophical point of view having a theory of everything is very attractive. It is neat to think there is a single, simple explanation behind everything. Nature suggests very strongly that it exists. But whether it actually matters? I don't think it does."

There is no answer?

Simply because we are very good at asking questions does not mean that we should assume there is always an answer. There may be no answer and we may just have to accept that.

I do not believe in God, certainly not as a separate entity that created the universe, and therefore do not believe that we are here to serve any purpose. We are just here and it makes no sense to ask why.

Understanding the universe around us will take time, a long time, if ever. But we are making progress and things are beginning to make sense. Most things are beginning to make sense. The one thing however that we will *never* be able to understand is how the universe came into existence. (See "Where did the universe come from?") How could it possibly come into existence? Where did it come from? What caused it to come into existence? The whole thing is just downright impossible, yet here we are, proving that it *is* possible. Annoying isn't it!

As I described in "Is there a reason for our existence?" we are all a part of the universe. It is not possible for us to examine the universe as if it were something separate from us, as if we were looking at it from the outside, and indeed this is the point that Stephen Hawking made. The very atoms that make us, the same atoms that were born in the fires of stars, are part of the fabric of the universe itself. We are as much a part of the universe as any region of 'deep space' or any remote galaxy. We *are* the universe. Our individual existence is a but a fleeting moment in the history of the universe, it is just a particular and brief arrangement of assorted atoms collected from all manner of places. Before we came into existence our atoms were arranged differently, forming other objects on our planet, or perhaps on other planets. Some of those atoms could have formed part of a giant Redwood tree, others perhaps the rocks under our feet. Before that period those same atoms could have been floating freely across space for millions of years as part of a massive molecular cloud. Before then the atoms would have been forged inside giant stars under extremes of pressure and temperature. Some of our atoms would have required even greater temperatures and energies to form and would have been created at the moment of a stars death, when it exploded in a massive super-nova. We could, theoretically, trace back the individual history of each one of the billions of atoms that are currently arranged to make us. It would be interesting to learn how many different objects we have been part of in the universe, and will be again when we die, and how far we have travelled across the universe since its incredible creation.

It is interesting to note that at the very moment of creation of the universe, the entire universe was contained inside an area smaller than one single atom. At that time there were no atoms, all was energy, and all was connected. As the universe has expanded and cooled matter was able to form. (See "Is the Big Bang Theory Correct?") Even though the universe is now many billions of light years across, it would appear that there still exists a connection between atoms that were once connected. (See "What is Quantum Theory?") This was discovered as a result of the EPR paradox. It is therefore quite possible that every one of us is - in some mysterious way - still connected, with not only everyone else, but *everything* else in the universe. We are just one huge, connected organism that has developed self-awareness. In other words we are all part of an intelligent self-aware universe.

So what am I? I am a collection of re-cycled atoms that is currently arranged in such a way that I exist as a conscious entity. When I die my atoms will once again be re-cycled into the universe at large and will form parts of other objects. One day in the future I may even be in the rain that falls onto my great grand children's garden. I suppose that is a kind of immortality.

In the meantime we may as well just sit back and enjoy the ride. God knows where it's taking us, but all will be revealed - one day. In the meantime perhaps we should delve deeper into the mysteries of quantum mechanics, relativity, light speed, religion, the Big Bang Theory, extraterrestrial life, time, time travel......but just a minute, I think this is where we started, but we still haven't properly resolved the problem of whether or not God exists, only that I do not believe in the God of the bible. I do though consider that our understanding of God is central to our understanding of ourselves, the universe and our place in it, and so have reserved that subject for the final topic with which to close this book.

25. Science, the Universe and God

"The only incomprehensible thing about the universe is that it is comprehensible."
Albert Einstein

There is something about the human race that separates us from the animals, or at least we believe there is, but is it true? It is difficult, if not impossible, to describe what that may be. I have argued that I do not believe it is our intelligence, and as an example I cited the case of those who are born with some form of brain damage and ask if that places them on the the level of an animal? Merely asking the question is enough to raise serious indignation all round, and quite rightly so. I have also said that I do not believe it is our appreciation of music, poetry, or any of the arts that makes us human, as I believe that is only as a result of our intelligence. I also do not believe it is our love for one another, or our desire to give help to others where needed without seeking personal gain - exhibiting altruism. Animals have been observed to show altruism. However, I dare say that expert naturalists will claim that when we observe say, a herd of elephants working together to rescue one of their young from drowning that they are only working for the benefit of the herd as a group. That being the case you could use the same argument for us humans.

We can say with confidence that animals exhibit love for one another, and even altruism, they will defend their young, are able to plan ahead and work together as a group. They exhibit joy and tenderness to their new-born and sadness and mourning at the loss of a loved one. So what do they not have that we do? I think the answer to that is imagination, and because we posses imagination we view the world very differently from the animals.

It is, I believe, because we have the power of imagination that we are able to marvel at the beauty of a mountain, or a sun set, or a wild flower, or a painting.

Imagination also affects the way in which people can relate to one another, and have the ability to completely understand how the other feels, sometimes without so much as word passing between them. All these attributes contribute to making us what we are, human, but to be human is far more than that. To be human is to posses understanding, to posses an awareness of our environment, and further more, to be human is to seek knowledge. Only a human being will look at a thing and ask 'why is that?'

When I peer through my telescope at a distant galaxy, or star cluster, or at countless thousands of stars in the Milky Way, I'm not just looking at tiny points of light or 'strange fuzzy blobs', I am looking at the universe, at our home, and it is stunningly beautiful.

I'm sure my dog, Sox, can see the Milky Way, but I'm also sure she only sees it as tiny spots of light and nothing more, if she notices the stars at all. I, on the other hand, see gigantic spiral galaxies that may harbour alien life, beautiful nebula where new stars are being formed, the remnants of an ancient supernova explosion created in a star's final death throes, the magnificent rings of Saturn and the moons of Jupiter. I see all these things and more, and marvel at them today just as I did as a 13 year old kid with his first look through an astronomical telescope. I'm still that 13 year old kid, he's never left. I find it impossible to look through an astronomical telescope and not be moved by the sight of countless stars in the night sky shining like diamond dust on a bed of black velvet. Multitudes of stars beyond stars, stretching back and back and finally becoming so many and so dense they appear to merge into delicate wisps of sparkling mist. I also find it impossible not to wonder where we, the human race, fit into 'the great scheme of things'. Are we here merely as observers, to watch as the universe evolves? Or are we here to participate, to shape the universe, to map out its - and our own - future?

Are we really here just by accident? Or are we at some deep level an essential and integral part of the universe without which there could be no universe? Have we to accept that having moved on from the early false belief that the Earth was at the centre of the universe, that we must travel full circle in our beliefs and accept that not only are we central to the universe, but that we *are* the universe?

I have said in this book that I do not believe in God, a God that created the universe, planet Earth or the human race. I have also said that I do not believe we, or the universe, are here for a reason, to fulfil a plan. I believe we are just here.

I will now elaborate on these points in order to reflect more clearly what I do believe.

I do not believe in God, the God as previously described here and in the bible, a God that created us, cares for us, watches over us and occasionally performs miracles for our benefit. No, that is a fairy tale God. That God, I am certain, is nothing more than a man made myth that is reinforced by the church as they struggle to keep power, and wealth. That God was invented by man to serve man's purpose. He was responsible for making crops grow, for the weather, the rising of the Sun and just about anything that we did not understand. In the past if a deadly disease spread through a population killing people in large numbers it was thought to be the work of God and prayers would be offered up asking for forgiveness of sins. Today we find it far more beneficial to be vaccinated against disease and have efficient sewerage systems and clean drinking water. As science has found explanations for what were in the past considered to be great mysteries, so the need to call upon God as a means of explanation has reduced. These days I cannot think of any serious phenomenon that we need to attribute to the mysterious workings of God, except perhaps the creation of the universe.

When we use the word 'God' we are referring to a supernatural being that possesses power that far transcends anything that we are capable of. We imagine God as having unlimited power, unlimited knowledge and having eternal existence. I find it strange therefore that God is often depicted in human form, as though He really is limited to a small physical shape. Perhaps this is because Jesus is recognised as being the son of God, and Jesus was, at least in appearance, human. I do not doubt that a man called Jesus did live a life very similar to how it is described in the bible, but believe that he was no more than a mere mortal the same as the rest of us. Was he the son of God? I do not think so for one minute, nor do I believe he performed miracles or rose from the dead. I am sure that Jesus was just a man, but a man that cleverly orchestrated events so that he appeared to be following the prophecies that foretold of the coming of the messiah.

The picture of God that we somehow have fixed in our heads as having a human shape, must surely be wrong, a relic from our childhood and all those cosy stories told in church. We know it is wrong yet it persists. Even if we manage to rid ourselves of this image of God we still seem unable not to form some sort of representative image of Him, whether it be in the form of a nebulous spirit or an inexplicable universal force. God has many faces.

No matter what image we prefer to use in our attempt to visualise or represent God, I am sure it is wrong, for we always seem to visualise God as existing *within* the universe. However, for God to be omnipotent and omniscient He surely cannot be constrained within the confines of the universe. He surely cannot be said to be even in any one place within the universe, He must surely be omnipresent. For God to be all of these things is impossible if we confine Him within the restraints of the universe, for how could an omnipotent, omniscient and omnipresent God possibly be any less than the universe in its entirety?

God can be nothing less than the entire universe. Every atom of it, every star, planet, galaxy, rock, person, creature, is literally a physical part of God, as God is embodied in all these things, every single atom. God is us, we are God, the universe is God. We are all a oneness, a part of the whole. It is impossible for us to examine God as a separate entity, just as it is impossible for us to examine the universe as a separate entity, for God, the universe, and ourselves, are all the one and same single entity, there are no divisions.

Does this mean that we are, after all, here for a reason? No, I think not, at least not in this context. However, the human race, will I believe, play a vital and pivotal role in the development of the universe, but not because that is the 'plan', but because that is what we do. We watch, we study, we experiment, we learn, we change things. It is perhaps the one thing the human race is good at, and in continuing to explore and learn, we will, I am sure, achieve wonderful things. At some future point we will surely gain mastery over the laws of physics, we will harness the awesome power of the universe, and eventually we will even understand the mysteries of time itself. When we have finally achieved mastery over time, we will then reach back through the ages and by our collective will bring about the creation of the universe. The universe was created because at some future time we would bring about its creation, but until we understand the terms 'future' and 'past' we are unable to make sense of it.

We do know however, that at the moment of creation of the Big Bang from a singularity the universe was contained in an area smaller than a single atom. At this microscopic level it is the laws of quantum theory that rule, and as we have already seen, in quantum theory particles not only interact instantaneously - thus at faster than light speed - they also have the ability to travel back and forth through time. We also know that at the moment of creation of the universe so time was created along with the normal dimensions of space. We also know that at the quantum level particles such as electrons and photons exist in a sort of phantom world of probabilities, until they are observed. Once observed they then adopt the reality of becoming a particle. It is this required act of observing to create reality that has led some notable theoretical physicists, such as John Wheeler and Stephen Hawking for example, to put forward an argument that it is only the presence of conscious observers, in the form of ourselves, that has

Keith Mayes

collapsed the wave function of the universe and made the universe exist.

Could it be that at the moment of creation of the universe, when time had no meaning, that events were put in place by ourselves, billions of years in the future, that would bring about the creation event by our conscious observations?

If there ever was, or ever will be, such a thing as a reason for our existence, then that is it, we are here to do what we have already done, to create the universe. After all, God can do anything, even bring about his own creation.

Bibliography

My sources are wide and varied and spread over a period covering many years, but much is from my personal book collection relating to cosmology, astronomy, space flight, physics and related subjects, which I have listed below.

'The Origin of the universe' John Barrow
'The Ascent Of Man' J. Bronowski
'Paradigms Lost' John L Casti
'A Man on the Moon' Andrew Chaikin
'Bubbles, voids and bumps in time: the new cosmology' James Cornell
The Alchemy of the Heavens' Ken Croswell
'Wrinkles in time' Davidson & Smoot
'The Mind of God' Paul Davies
'Kinds Of Minds' Daniel C. Dennett
'The World Treasury of Physics, Astronomy, and Mathematics' Timothy Ferris
'The Whole Shebang' Timothy Ferris
'Six Not-So-Easy Pieces' Richard Feynman
'QED The strange theory of light and matter' Richard Feynman
'Key Studies in Psychology' Gross R
'Sociology' Flemming M. Hine B. Nobbs J
'One Giant Leap. The extraordinary story of the moon landings' Tim Furness
'Alice in Quantumland' Robert Gilmore
'The elegant universe' Brian Greene
'Companion to the Cosmos' John Gribbin
'In search of Schrodinger's cat' John Gribbin
'Schrodinger's kittens and the search for reality' John Gribbin
'Blinded by the light' John Gribbin
'Sociology Themes and Perspectives' Haralambos and Holborn
'A Brief History of Time' Stephen Hawking

'The Universe in a Nutshell' Stephen Hawking
'Hyperspace' Michio Kaku
'The nature of space and time' Stephen Hawking and Roger Penrose
'The Emperor's New Mind' Roger Penrose
'The Large, the Small and the Human Mind' Roger Penrose
'The Mathematical Tourist' Ivars Peterson
'The shadows of creation' Michael Riordan and David Schramm
'The Holy Bible' King James Version

About the Author

Keith Mayes has been a keen amateur astronomer for the past 45 years and will talk about his favourite subject to anybody who cares to listen. It is his fascination with the distant stars and galaxies that led Keith to the study of cosmology and physics.

Another of Keith's keen interests is photography, so it is no surprise that he has linked this with astronomy and developed a passionate interest in astrophotography. Some examples of his work in this field can be seen on his web site at www.thekeyboard.org.uk. and one of his deep sky photographs of the famous Orion Nebula appears on the jacket of the book 'Alien Physics' by Ronald Avery, also available from www.1stbooks.com

Keith was born in London in 1946 but moved to Scotland when he married his Scottish wife in 1990. They live in a small town on the west coast of Scotland and share their house with their cat and dog and Keith's 8 inch Schmidt-Cassegrain astronomical telescope. Keith has always had an enquiring mind and is an avid reader of all things scientific.

He has spent many years researching cosmology, astronomy and physics and found much of interest in such diverse subjects as quantum theory, artificial intelligence and relativity.

Keith has spent a lifetime pondering over the mysteries of life and the universe, and this book is the result.

Lightning Source UK Ltd.
Milton Keynes UK
173812UK00001B/3/A